CONSTRUCTING DESTRUCTION

CRITICAL CULTURAL HERITAGE SERIES

Series Editor: Beverley Butler
Part of the University College London Institute of Archaeology Publications Series, published for the Institute by Routledge.
General Series Editor: Ruth Whitehouse
Founding Series Editor: Peter Ucko

The aim of this Critical Cultural Heritage Series is to define a new area of research and to produce a set of volumes that make a radical break with the existing canon of cultural heritage texts. In a fundamental shift of perspective, Jacques Derrida's rallying call to "restore heritage to dignity" inspires both a reexamination of the core question of what constitutes cultural heritage and an engagement with the ethical issues that shape the possible futures of this research area.

The series is intended to be of transformative value in creating new agendas within cultural heritage discourse, using individual texts as building blocks. Central to the project is a realignment of cultural heritage studies through a wider scholarship committed to disrupting the Eurocentrism which underpins current theory and practice and through a contemporary "politics of recognition" that is concerned with articulating new, alternative, or parallel characterisations of heritage value. The aim is to centre cultural heritage studies within a wider concern for the preservation of human dignity and justice and to use these alternative discourses as a resource for future action, thereby creating a proactive, responsive, and just future for both cultural heritage studies and heritage practice.

Volume 12: Trinidad Rico, *Constructing Destruction: Heritage Narratives in the Tsunami City*
Volume 11: Anna Källén, *Stones Standing: Archaeology, Colonialism, and Ecotourism in Northern Laos*
Volume 10: Pierre Lemonnier, *Mundane Objects: Materiality and Non-Verbal Communication*
Volume 9: Shaila Bhatti, *Translating Museums: A Counterhistory of South Asian Museology*
Volume 8: Marilena Alivizatou, *Intangible Heritage and the Museum: New Perspectives on Cultural Preservation*
Volume 7: Charlotte L. Joy, *The Politics of Heritage Management in Mali: From UNESCO to Djenné*
Volume 6: Layla Renshaw, *Exhuming Loss: Memory, Materiality, and Mass Graves of the Spanish Civil War*
Volume 5: Katharina Schramm, *African Homecoming: Pan-African Ideology and Contested Heritage*
Volume 4: Mingming Wang, *Empire and Local Worlds: A Chinese Model of Long-Term Historical Anthropology*
Volume 3: Dean Sully, Ed., *Decolonizing Conservation: Caring for Maori Meeting Houses outside New Zealand*
Volume 2: Ferdinand de Jong and Michael Rowlands, Eds., *Reclaiming Heritage: Alternative Imaginaries of Memory in West Africa*
Volume 1: Beverley Butler, *Return to Alexandria: An Ethnography of Cultural Heritage Revivalism and Museum Memory*

CONSTRUCTING DESTRUCTION

Heritage Narratives in the Tsunami City

Trinidad Rico

Routledge
Taylor & Francis Group
NEW YORK AND LONDON

First published 2016
by Routledge
711 Third Avenue, New York, NY 10017

and by Routledge
2 Park Square, Milton Park, Abingdon, Oxon OX14 4RN

First issued in paperback 2018

Routledge is an imprint of the Taylor & Francis Group, an informa business

Library of Congress Cataloging-in-Publication Data:
Names: Rico, Trinidad, author.
Title: Constructing destruction : heritage narratives in the tsunami city / Trinidad Rico.
Description: Walnut Creek, California : Left Coast Press, Inc., [2016] | Series: Critical cultural heritage series ; volume 12 | Includes bibliographical references and index.
Identifiers: LCCN 2015042680| ISBN 9781629584379 (hardback : alk. paper) | ISBN 9781315520933 (eBook)
Subjects: LCSH: Cultural property—Indonesia—Nanggroe Aceh Darussalam. | Nanggroe Aceh Darussalam (Indonesia)—Antiquities. | Nanggroe Aceh Darussalam (Indonesia)—Cultural policy. | Tsunami relief—Indonesia—Nanggroe Aceh Darussalam. | Ethnology—Indonesia—Nanggroe Aceh Darussalam.
Classification: LCC DS646.15.A8 R53 2016 | DDC 959.8/11—dc23
LC record available at http://lccn.loc.gov/2015042680

ISBN 13: 978-1-138-58037-4 (pbk)
ISBN 13: 978-1-62958-437-9 (hbk)

Typeset in Times New Roman
by Ryan Kenney

To Michael and Nahuel

Contents

List of Illustrations

Preface

In the early planning stages of this research project I was issued a collegiate warning about embarking on research in and about Aceh: there is not a tremendous amount of local knowledge about or even interest in what we could call "tangible" cultural heritage in Aceh. Believing this to be inadequate, and possibly unwarranted, this warning became a personal call to arms to interrogate the practices that shape a heritage landscape and its interlocutors in post-tsunami Banda Aceh and its environs. The resulting research project sought to document vernacularized constructions of post-tsunami heritage and a local preservation philosophy characterized by subtleness and therefore invisible to many, hinging on the identification and documentation of destruction and construction of heritage subjects and practices. But the intersection of post-tsunami devastation and the search for alternatives to already established practices and heritage concerns that were perceived to be harmless was difficult to convey, making communication difficult with fellow disciplinary colleagues in Aceh and elsewhere in Indonesia. I interpreted this as a sign of a shortage of critical debates on heritage practices in the region at the time, and navigated it by explaining often that, when it came to my research efforts, Indonesia *adalah tempat yang sama bagusnya dengan tempat lain* (Indonesia is as good a place to work as anywhere else). This volume argues that nothing could be further from the truth.

Notes from "the Aftermath"

The fieldwork that informs this volume was carried out between 2007 and 2010, at a time when the characteristics of the "new" city of Banda Aceh were being negotiated and manifested following the devastating 2004 Indian Ocean tsunami. With a strong focus on desirable futures, the timing of this research project was best described as being about "the aftermath." The tsunami catastrophe, the Helsinki peace negotiations, and the peak of reconstructions had already taken place. However, the project was chronologically braced by two transnational incidents that characterized the tone of heritage debates in the wider region. At a global level, the final stages of the New 7 Wonders of the World competition were taking place in mid-2007. While the announcement was to take place in Portugal, Southeast Asia took the voting to heart. Malls in Bangkok were lavishly decorated with architectural models of the heritage sites that had made the final round of selection in this competition, and retail shops offered "New 7 Wonders of the World" discounts. It was a New 7 Wonders of the World extravaganza, showing overwhelming support for a campaign that was launched in 1999 by the Swiss adventurer Bernard Weber. Later in 2007, the contest caused significant disappointment in Southeast Asia when the final seven sites selected failed to include Angkor Wat in Cambodia or any other Asian heritage site. Across different social media platforms, Southeast Asians asked what could have gone wrong in the online voting system as evidenced in a July 9 editorial in *The Nation* adding that, "for serious-minded people, the safest bet to preserve the world's cultural heritage is to have them listed as World Heritage Sites." Despite presenting itself as the most democratic and accessible assessment of global heritage significance that relied on an online platform for dissemination and voting, the New 7 Wonders of the World initiative was no match for the popularity of the World Heritage List hosted by the United Nations Educational Scientific and Cultural Organization (UNESCO), whose brand and expertise enjoyed dominance in the process of heritage legitimization. None of the heritage sites of Indonesia had been selected at the final stages of this contest. However, the sequel, the New 7 Wonders of Nature, pleased Indonesians in late 2011 with the successful inclusion of Komodo National Park in the final selection, joining other winning natural wonders in Vietnam (Ha Long Bay), South Korea (Jeju Island), and the Philippines (PP Underground River). Four of the seven winning sites in this contest were located in East and Southeast Asia, a reminder that, at the global scale, natural heritage is a more marketable resource than cultural heritage in this region of the world.

At a national level, another process of heritage legitimization was taking place. In 2007, Malaysia launched a tourism campaign with the controversial slogan "Truly Asia," which was criticized all over Asia for its claims to regional authenticity. Addressing the use of what were then perceived to be traditional Indonesian songs and dances in this campaign, Malaysia and Indonesia signed an agreement to honor each other's cultural heritage. But in 2009, Indonesia was outraged by the use of imagery from the Balinese *pendet* dance in a promotional video for Malaysia, an act interpreted as a cultural claim in which

Malaysia was taking over Indonesian traditions (see Chong 2012). Existing cultural tensions escalated: Malaysian flags were burnt in Indonesia, the Malaysian embassy was egged, and rallies were staged by Balinese artists. The minister for culture and tourism for Indonesia, Jero Wacik, condemned the exploitation of the *pendet* dance without permission from the Indonesian government, stating, "The entire world knows the *pendet* dance belongs to Bali. So it is a must for Malaysia to get the permits first from Indonesia before using it" (Simamora 2009). In his view, they had violated their 2007 agreement. The matter was resolved only when blame could be assigned to the producers of the commercial that was aired worldwide. Discovery Networks Asia-Pacific issued a statement assuming responsibility and regretting the wrongful use of this footage, which was sourced from an independent third party. Discovery Channel then produced a different program on Indonesia's traditional Bali dance, to ease the tensions between Malaysia and Indonesia. Meanwhile, Malaysia's home affairs minister, Hishammuddin Hussein, assured the safety of 1.8 million Indonesians in Malaysia, in spite of the provocation, pressure, humiliation, and threats that they had received. The question of transboundary cultural legitimacy at the time was excessively central to the heritage concerns of some heritage circles across Indonesia during the time of my fieldwork, especially articulated through conversations of the proper use of UNESCO's Convention for the Safeguarding of the Intangible Cultural Heritage. A patriotic resistance to products of a hybrid cultural narrative between Indonesia and Malaysia had manifested at multiple levels and through a variety of geopolitical issues as well as politicized cultural and social clashes (Clark and Pietsch 2014). In this particular case of cultural confrontation, a concerned heritage public referenced instruments of global governance, such as UNESCO's Representative List of the Intangible Cultural Heritage of Humanity, or international copyright treaties, such as the Berne Convention, to support their claims to heritage ownership.

Conversations about tangible heritage and intangible heritage matter in Southeast Asia (Byrne 2014; Winter and Daly 2012). Yet, however dominant these timely tensions appeared to be in regional heritage debates, questions of inclusion in or exclusion from a coherent heritage assemblage did not seem to hold the same relevance in Aceh. Accustomed to a peripheral role in Indonesia's monumental heritage, it could be argued that the Acehnese did not care for the validation of their cultural heritage at the global monumental level that the New 7 Wonders of the World was promoting. Moreover, it could be argued that Aceh remained more in tune with the identification of cultural ties with Malaysia and the associated overlap in cultural performances and traditions due to its historical role as a productive center of Malay-Islamic literary activity and as an Islamic model in Malay courts (Clark and Pietsch 2014, 117). During this same period, cultural debates in Aceh should be seen instead as preoccupied with reformulating its distinctive regional identity under the close supervision and influence of the international community, the Indonesian state, and the expectation of cultural experts local and foreign alike. These processes would impact the shape and nature of its heritage assemblage, and inspire the research that gave rise to the present volume.

In line with my disciplinary position to document and discuss a self-determined construction of heritage that is able to shift contextually and temporally, the Indonesian spelling throughout this book does not follow any particular convention. Instead, I use language in a way that reflects how most of my informants and colleagues used spelling in their own work, through the notes that they shared with me, and the corrections that they made to my written work.

Acknowledgments

Research for this volume was completed thanks to the generous academic and financial support of Stanford University, specifically the following departments: the Department of Anthropology, the Stanford Archaeology Center, the Abbasi Program in Islamic Studies at Stanford, the Center for Philanthropy and Civil Society (PACS), and the Office of Community Engagement. Additional stages of writing were supported by UCL Qatar and Texas A&M University at Qatar. The completion of this volume should also be attributed to many mentors who contributed to this research project with their guidance and expertise: Lynn Meskell, Ian Hodder, Paulla Ebron, and Denis Byrne provided invaluable feedback since the early days of fieldwork and writing, and commented on earlier versions of some of the chapters that make up this book, while Rodney Harrison was instrumental in the shape of the final draft. Possibly unbeknownst to them, Anna L. Tsing, R. Michael Feener, and Shahzad Bashir also contributed informally to the shaping of this volume through their inspirational work and availability for conversations. Shelly Coughlan and Ellen Christensen at Stanford University, Simona Duranti at UCL Qatar, and Hamza Bin Jehangir and Giulia El Dardiry at Texas A&M University at Qatar supported my work logistically at different stages of production. Finally, there are countless people in Indonesia to thank for their patience and support, but in particular, the friendship and generosity of my Indonesian host families in Aceh—Yeyen's family in Banda Aceh and Eddy's family in Lhok Nga—made this project worthwhile above all else.

Introduction: The Problem with Destruction

The future value is just as relevant.
—Frits van Voorden (1999)

In 2007, the city of Banda Aceh could be aptly described as a destroyed landscape following the late 2004 Indian Ocean tsunami. On the west bank of the mouth of the Kreung Aceh River in Gampông Pandé, expert architect and preservationist Dr. K. introduced me to the city for the first time. It was, incidentally, his city—as a native of Aceh, transplanted to Java through studies in Europe, he continued to be involved in Aceh through his work. The National Committee of the International Council on Monuments and Sites (ICOMOS) in Indonesia had put us in contact, so we arranged to meet at the start of this project, flying from Jakarta together. Considering the circumstances of our introduction, my expectation was for this visit to focus on the typical foci that are of concern to such organizations—namely, a tour of colonial structures and other monumental heritage places. But I was pleasantly surprised. At the river mouth on Gampông Pandé, the landscape was filled with tsunami trash and debris, which was still being washed ashore and deposited along the coast. The floor was scattered with a significant density of potsherds of a type that I later had identified informally at first sight as "generic Chinese" pottery.[1] There were no structures except for an abandoned wooden shop house and a dilapidated fishing boat resting on the ground. Further away, the desolation seemed to continue, without a "heritage site" in sight. Nonetheless, my host spent a good hour on this spot, describing for me an invisible sultan's fortress whose remains may or may not exist under the unexcavated soil. This conversation would set the tone for the visit and the way that heritage was discussed from here on. It was clear that, for my colleague, heritage in Aceh was about *construction*; his was a future-oriented perspective full of possibilities, as if much of the heritage of Aceh consisted in the fragments of a broken city that were on the brink of being reassembled for this purpose.

On this very spot a plaque on the side of the road read, in Indonesian, Acehnese, and English, "This mark[s] the origin of Kota Banda Aceh, where Sultan Johansyah established the Sultanate of Aceh Darussalam on the 1st of Ramadhan 601 H (April 22, 1205 AD)."

This was not purely serendipitous. The authorship of the plaque was attributed to the Pustaka Bustanussalatin Foundation, an organization in which Dr. K. held a prominent role. Over the following years, plaques like this were installed across Banda Aceh, commemorating events from the past and bringing historical significance to urban and natural landscapes. Only some of these plaques accompanied the monuments whose chronologies and historical contexts they paid homage to. Instead, many of these described events that were less prominent in existing historical and heritage narratives and barely tangible within the landscape. This was perhaps the first indication that the heritage of Banda Aceh was not destroyed after the tsunami in late 2004, but, rather, heritage had a key role to play in the reconstruction of the city—informed and configured by the events of the past, and enhanced by a unique approach to present and future imaginations of the region.

This proposition was already materialized in a new heritage entity with growing visibility, the "tsunami boat." As it could be expected from the nature of the tsunami disaster, there were a large number of boats that were relocated by the tsunami forces to different places inland, but only a handful of these became established in the landscape and contemporary narratives of Aceh as the city was subjected to intense reconstruction. Tsunami boats are not unique to Aceh. They are prominent symbols of the aftermath of the Indian Ocean disaster across South and Southeast Asia. In Thailand, for example, the "Police Boat Memorial" in Khao Lak and the "Tsunami Boat Memorial" in Baan Nam Kaem are preserved in situ, marking the areas in the country that were hit the hardest by the tsunami. These boats as curios have been widely circulated as symbols of postdisaster landscapes, representing the displacement and chaos that took place during the catastrophe. As objects out-of-water they add poetry to postdisaster disorder: identifiable debris with a hint of irony, a logistical problem of gigantic proportions to be admired, and a powerful image that suggests yet manages to avoid direct contact with human loss. Tsunami boats also refer to a shared traditional livelihood in the region that became endangered, a common denominator that could transcend cultural and geographic boundaries, much like the disaster did in late 2004.

In Banda Aceh, this process of relatively random memorialization turned an ordinary fishing boat and an industrial electrical ship from ruins into relics, as they went from their moorings in Ulee Lheue harbor and the Kreung Aceh River to rest at various locations in Banda Aceh. Intersecting the significance of the ocean in the history and livelihood of Aceh and the medium of an ocean-borne natural disaster, it could be said that the boat was the perfect candidate for the centerpiece of commemoration and postdisaster self-representation. Many dislocated vessels had been removed as soon as it was logistically possible, seeing that they interfered with reconstruction plans or circulation along key areas of the city, and of these only photographs remain. For example, the boat that landed at the entrance of Medan Hotel was removed in 2005 for blocking one of the entrances to the military base in the center of town. Initially facing a similar fate, the fishing boat in Lampulo and the electrical boat named PLTD Apung 1 in the area called Punge Blang Cut were instead the focus of a process of heritagization over the same period (Rico 2014). These boats gradually came to stand out, acquiring significance and avoiding the state of ruin through ongoing and moderately complicated efforts of stabilization that fall firmly within recognized practices of heritage preservation. As I will discuss in Chapter 1, significant efforts were made toward improving the survival and integrity of these new monuments, efforts that would suggest that these were, in fact, objects of heritage value that require safeguarding and care. However, as I have discussed elsewhere (Rico 2014), "heritage," as it is commonly understood, is not a construct that is readily capable of capturing and protecting this type of contemporary significance. A dominant heritage framework operating on global and local scales is concerned instead with strict chronological boundaries that give legitimacy to material and discursive traces of a more distant past. Tsunami boats, as the embodiments of recent events and experiences (despite their strong potential to become temporal markers of a historical event in due time), do not fit this ideal. The dominant framework imagines, above all, a heritage that is endangered, chronologically remote, and in need of safeguarding. Therefore, a disciplinary focus on establishing and acting upon perceptions of traditional "heritage at risk" leaves tsunami boats and other unorthodox heritage constructs in Aceh unsupported by the dominant heritage framework. As a result, this marginalization could result in an eventual erasure.

From Destruction to Construction

There is an intimate relationship between destruction, risk, and heritage. The fear of decay, destruction, and loss lies at the heart of all constructions of heritage. This fear is operationalized by way of a series of activities, staged as active interventions that are intended to curb, or at least minimize, the risk of such decay or destruction. These forms of heritage protection, conservation, preservation, and management operate at a range of different spatial scales and are distributed among the actions of a range of heterogeneous amateur and professional bodies—local, regional, national, and international. Their operation is underpinned by a number of different apparatuses with which various forms of "risk" and "endangerment" are quantified, visualized, and made measurable. All heritage is thus, by definition, somehow "at risk," whether implicitly (because of its age, rarity, uniqueness, value, or other unquantifiable threat) or explicitly. While the central place of loss and decay to heritage has long been recognized, the ways in which notions of risk and endangerment are operationalized and gain strength as a set of material-discursive practices that naturalize and enable the depoliticization of heritage management have only recently begun to be recognized (e.g., Harrison 2013; Meskell 2014; Vidal and Dias 2016). This rhetorical concept of "heritage at risk" produces a set of circumstances in which practices of heritage conservation and management are understood to be "natural" and morally incontestable, justifying approaches that suggest a universal concern for conservation and its priorities (Rico 2014, 2015).

Destruction, decay, and ruination—as forms of change understood to be universally negatively framed—must be recognized as key concepts that discursively operationalize forms of heritage management that in reality serve a variety of sociocultural, historical, and political agendas. Heritage in postconflict and postdisaster scenarios assumes special significance as a subject of study for these mobilizations, as it comes to act as a powerful material witness to the perceived vulnerability of all heritage and a parable of the threats that would overcome it in the absence of active management and intervention. Such parables, of flooded, bombed, and destroyed monuments, have been put to work to justify a series of international safeguarding campaigns, the World Heritage Convention, and various other national and international instruments concerned with the preservation of heritage in all of its increasingly expanding forms: tangible and intangible, monumental and vernacular, rare and representative. The ways in which heritage is managed within the contexts of such disasters, with the subsequent circulation of images and texts relating to the destruction of heritage these disasters have caused, reveal the operationalization and manipulation of a broader heritage "endangerment sensibility" (cf. Vidal and Dias 2015) at a range of different spatial and discursive scales.

If we are to understand "heritage at risk" as a rhetorical device that justifies and facilitates various forms of material and discursive interventions that address specific sociohistorical and political ends under the guise of universal values, a central discussion throughout this volume is the way that the construction of heritage is intimately associated with processes of physical and discursive destruction that are mediated and selective. In this sense, those landscapes that I refer to as "postdestruction" landscapes (i.e., postconflict and/or postdisaster landscapes) are presented as critical contexts for revealing the forces, categories, and agencies that shape heritage constructs.

The ontological question that frames the case study presented here is whether heritage can be constructed independently of its actual or potential destruction. This question invites ethnographic approaches that champion the specificities of locality and chronology in the construction of destruction as a referent in heritage discourses and practices, contributing to our understanding of situated perceptions of value, tangibility, authenticity, time, and the construction and imagination of catastrophe. In doing so, local cosmologies are sought with the objective of understanding whether heritage may exist without a need to be preserved at all.

Such contexts are not only vital to an understanding of the production of heritage value in their guise as the negative parables that justify active practices and processes of heritage management elsewhere. There is another important way in which postdestruction landscapes are significant to the study of heritage and change. Postdestruction landscapes constitute fertile and active grounds for the production of *new* forms of heritage. They do this in three interconnected ways. In the first instance, postdestruction landscapes contain material scars and absences that are made to speak to the survival of destroyed or lost heritages that may themselves come to act as memorials in a landscape of loss. A good example of this is the way in which the empty niches of the Bamiyan Buddhas could be said to have come to acquire as much if not greater significance in their absence than they may have had if they had not been destroyed in the first place (see further discussion in Harrison 2013, 182–192). Secondly, postdestruction landscapes often contain new material and archival traces that might be made to speak to the heritage of the circumstances of the destruction that come to acquire value in and of themselves. For example, Andrew Herscher (2010) has shown how traces of violence against architecture in the Kosovo war must be read historically within the context of a longer trajectory in which identities are continually shifting and emergent in relation to selective processes of destruction and preservation of architectural monuments. And thirdly, disasters often clear the way for the production of new forms of heritage that were not previously able to be memorialized in a landscape through processes of material loss and an associated unburdening from previous memorials and constructions of the past. The classic example of this process is iconoclasm, where one set of religious, political, or cultural symbols are cleared away for the production of new ones. However, the value in studying a postdisaster as opposed to a postconflict landscape in this regard is that it allows such processes to be studied independently of the requirement for moral accountability, which is often held against the work of iconoclasts. Thus, rather than simply alluding to change, an investigation of the postdestruction landscape reveals an ecosystem of production of heritage value that operates beyond an extinction framework (Lowenthal 1999).

What is at stake in the recognition of the ways in which risk, endangerment, and destruction come to figure in the production and operationalization of heritage in postdisaster landscapes? My aim is to question the operational limitations of conceptualizing heritage within a framework dominated by threat assessment and management, and to challenge the validity of blanket "at risk" categorizations that are not constructed locally. This book therefore considers how risk assessment may be irreconcilable with local constructions of cultural heritage that are rooted in specific understandings of destruction, decay, and impermanence, which would require a redefinition of notions of preservation, conservation, decay, destruction, and the various categories by which risks to heritage are quantified and made measurable altogether. An unquestioned embrace of a lingua franca driven by risk, decay, and destruction is potentially anathema to emerging heritage discourses in a postdisaster redescription of heritage value that may deviate from established sets of concepts, categories, and priorities, in which the emphasis on risk and destruction is key. I argue, for example, that an uncritical deployment of a "heritage at risk" rationale served to marginalize the emergence of places of heritage value in the process of legitimization of the reconstructed cultural landscapes of Banda Aceh. An emphasis on heritage that is perceived to be at risk may prevent the valorization and protection of locally grown heritage incarnations that do not comply with established global understandings of heritage constructs.

Structure of This Volume

Chapter 1 in this book, "*Khas Aceh*," defines Banda Aceh and the Aceh province as the place of productive enquiry into the construction and destruction of heritage. The rationale for examining Aceh and its heritage construction processes is embedded in the geographical location and contemporary situation of Aceh, highlighting an emphasis on uniqueness as a self-identifier throughout the past and recent history of this region. This introductory chapter also aims to describe and depart from a dominant historical engagement with the landscape and narratives of Aceh that rests authority on exogenous texts, scholarships, and foreign agendas. Instead, the method introduced in this chapter aims to position accounts in a postdisaster and postcolonial built landscape "in progress," resisting historically established trajectories and assimilating heritage values as values that exist in flux. The process of assimilation and countering of destruction is discussed as a way to recenter and localize the historical narrative so that ideas of Aceh and being Acehnese are rendered coherent and unique, a focus that has been subject to censorship in this separatist region.

Chapter 2, "Heritage Narratives in the Tsunami City," examines the relationship between history and a historical construction of the heritage landscape in the process of Aceh forging a new identity as a tsunami city during this period. Its focus is on the ways in which heritage and history coconstruct place. Heritage makes and continuously remakes a sense of place but not without navigating its historical baggage: history is mobilized and remixed in strategic ways. Echoing the subtitle of this volume, this chapter suggests ways in which heritage constructs are mobilized in two different types of intertwined narratives. On one hand, the management of heritage value can be seen as a practice that perpetuates, resists, or adapts to the forces of destruction. This is a resistance that is inherent to the idea of heritage, and embedded in established global, regional, and local heritage discourses and practices that are summoned in the work of reconstruction. On the other hand, locally emergent heritage constructs that are left to operate independently of aforementioned agendas may become stewards to memories and narratives that explain, interpret, and preserve collective experiences of destruction, catastrophe, and decay in the context of natural disasters. Heritage is, in this sense, a vehicle for the expression of a unique relationship between the Acehnese and their environment that derives from a recent history in process of being "materialized" through the commemoration of prehistorical, historical, and posthistorical vignettes in its landscape.

A historical and theoretically informed review of the way destruction has been centrally featured in the rise of heritage discourse and rhetoric is at the center of Chapter 3, "The Construction of Destruction." This chapter describes structures in heritage thinking that established the discipline at the global and regional level, also defining a heritage industry for Asia in contrast to a Western paradigm that is predictable but nonetheless effective in the shaping of this vast region. This chapter aims to demonstrate the way that an articulation of a safeguarding mission that is epistemologically central to the heritage discipline is mobilized rhetorically through perceptions of a "heritage at risk." This rhetoric naturalizes a rejection of destruction and empowers expert networks through the promotion of this concern and the strategies devised to address it. Destruction and its perception are also mobilized to recognize and categorize non-Western heritage constructs, a potentially Orientalizing practice that dichotomizes heritage between an Eastern versus a Western product that imagines a uniquely Asian understanding of destruction. The construction of heritage in Banda Aceh is presumed to be built on this conceptual territory, while in reality there is an expert discourse that mediates languages and knowledges that are legitimately Asian and Western at the same time.

The fourth chapter, "An Ethnography of 'Heritage at Risk,'" examines these negotiations in practice, through the lens of an ethnographically sensitive Acehnese construction of heritage that considers three themes that intersect in the construction of a post-tsunami heritage: the influence of the formalization of a "culture of disaster," the Islamization of tangible cultural heritage places and memorials, and the significance of a chronological sensitivity in the construction of heritage subjects. These localized descriptions demonstrate that heritage constructs effectively mediate and resist the victimization of culture, taking an active formative role in the construction of futures that are locally sketched to transmit disaster narratives into the future. Incidentally, these also enrich the resilience capability of Banda Aceh in accordance with existing cultural and religious frameworks that give local meanings to disaster and destruction. Building on the insights gained from an ethnographic methodology, the value that is ascribed to Acehnese heritage through collective memories, erasures, reappraisal, and resistances reveals how heritage construction corresponds with a categorical reorganization that follows natural disasters (Hoffman and Oliver-Smith 2002; Oliver-Smith and Hoffman 1999).

The concluding chapter in this book, Chapter 5, "Destruction Alternatives," brings together the different themes explored throughout the book to suggest an approach to alterity in heritage studies that focuses on the vernacularization of heritage constructs, and in particular, those that embrace destruction as a legitimate constructive practice. This discussion is situated in an era of heritage studies that encourages a holistic examination of local—and creative—adaptations of heritage standards and objectives with a strong focus on processes of vernacularization, as well as the web of institutions, expert networks, local and foreign agents, and media who transmit a destruction-based model of heritage construction and management through agendas, practices of list making, boundary setting, and other material and rhetorical devices.

The field of heritage ethnography is still relatively new (for example, Byrne 2007; Meskell 2011; Joy 2012). Heritage studies, through the documentation of heritage constructs and their preservation, is increasingly recognized as being about people, despite its disproportionate focus on objects, monuments, and their tangibility. New disciplinary approaches have emerged to capture a sensitivity or awareness of alternative worldviews and vernacular understandings and constructions of heritage. Early ethnographic heritage value assessments (i.e., Low 1981) were developed to reveal cultural representations in heritage experiences but have been criticized for lacking depth of cultural understanding (Ervin 1997). Improved approaches attempted to circumvent the problem of imposing established heritage categories and typologies, which interfere with cultural interpretations of localized constructs (Low 2002). Therefore, a decolonization of heritage preservation practices has aimed to involve participatory conservation models (Clavir 2001; Sully 2008; Warrack 2011) in order to acknowledge and document the way that values are coconstructed, negotiated, and shared among experts and marginalized groups (Clavir 2002). Ethnographically informed studies of heritage preservation go beyond an acknowledgment of the existence and rights of stakeholders, and require that the process of stakeholder consultation be refined and redefined to recognize, validate, and include marginalized voices and landscapes at every step of heritage management. In addition, this approach recognizes that indigenous cosmologies are not only a source of local perceptions of territoriality. They also constitute moral discourses, inculcating cultural values regarded as intrinsic to the customary order of things and people (Pannell 2007). Practical methods of cultural mapping (e.g., Byrne and Nugent 2004; Harrison 2004) have enriched the work of heritage, designed and put into practice to attach significance to the landscape as it is imprinted with life stories, histories, and emotions. Maps of attachment are thus constructed and recorded ethnographically to help bring greater visibility to the lives and histories of marginalized groups, translating, as landscape characterization does, indigenous category systems that reveal the intrinsic relationship between groups and the construction of culturally meaningful landscapes, which is essential to the field of anthropology (i.e., Tsing 2005; Kwon 2008). In this ethnography of heritage at risk, I propose to do exactly that.

1 Khas Aceh

Aceh as a fieldsite has been shaped by plenty of academic examinations, primarily through the lens of history, politics and, perhaps after the 2004 disaster, historical ecology. It is through this academic lens that Aceh as a place is made to unfold geopolitically, informing the channels of articulation of an identity for the city of Banda Aceh and its inhabitants that is invariably rooted in long-term processes. In this chapter, however, I aim to juxtapose the historical landscape and narrative of Aceh, which are created through this lens, with emerging counter-landscapes and counter-narratives. Since these forms of resistance are communicated through post-tsunami heritage discourses, they highlight the value of carrying out ethnographically informed reassessments of contemporary identities through the perspective of cultural heritage.

The Aceh province is the northernmost province of the Republic of Indonesia on the island of Sumatra. The province includes the much smaller islands of Sabang to the north, Simeulue and Aceh Singkil to the southwest. The capital city of Aceh, Banda Aceh, is set in the regency of *Aceh Besar* (Greater Aceh). Both urban and remote, it is a city that boasts a complex relationship to the Republic of Indonesia, to the Southeast Asian archipelago, to the Indian Ocean, and to the Islamic World, yet cannot be adequately characterized from any of these vantage points in isolation. Describing the region culturally, Holly Smith cites an acronym that represents this diversity: "Arabic, Chinese, European, and Hindu" (ACEH), identities that she argues are blended and unified within the art and culture of Aceh (H. Smith 1998, xii). Nonetheless, the homogeneity of this cultural diversity has been questioned. Instead, Aceh has been presented as a composite of culturally diverse areas (Kartomi 1981) influenced by external cultural connections (Ward 1975). However, from the perspective of the Acehnese, there is a uniqueness that goes beyond its cosmopolitan legacy. A commonly circulated phrase, *khas Aceh* ("Aceh-specific"), is widely used to refer to Acehnese distinction and that which is perceived to be traditional. The Acehnese traditional *rencong* is *khas Aceh*, and so is the delectable Aceh coffee. But ordinary bananas grown in Aceh may also be identified as *khas Aceh*, a qualifier that may be applied equally to the region's mangroves, local histories, and the wise gaze of the Acehnese themselves, unmistakable due to its likeness to "that of a bird of prey," as a proud Acehnese employee of the Badan Rehabilitasi dan Rekonstruksi (BRR) explained to me cultural identifier on the very first day I arrived in the city. *Khas Aceh* aptly summarizes the cultural phenomenon that is Aceh, and as a concept, it aids in the examination of the construction of distinctively Acehnese approaches to cultural heritage, both historically and in the present.

The latest chapter in the history of Aceh, the 2004 tsunami disaster, was unquestionably unique. On December 26, 2004, at 00:58 UTC, an undersea earthquake originated at an epicenter located between the island of Simeulue and mainland Sumatra, Indonesia, 150 km off the coast of Aceh (Nalbant et al. 2005). For the years following the catastrophe, a series of scientific sound bites were circulated globally when referring to this event: the entire planet vibrated, triggering other earthquakes as far away as Alaska; the shift of mass and energy slightly altered the rotation of the Earth, theoretically shortening the length of a day by a few microseconds; and the Earth wobbled on its axis. Some scientists explained that the energy released at the moment of the quake was estimated to be equivalent to over 1,502 Hiroshima atomic bombs. The earthquake was identified as having one of the longest durations of any earthquake ever recorded in history (Lay et al. 2005), registering at the incredible magnitude of 9.1 to 9.3 on the Richter scale. However, this impressive figure soon paled in comparison to the destruction caused by the tsunamis that the subaquatic earthquake triggered around the Indian Ocean, and so the disaster is more commonly referred to as a tsunami disaster.

The disaster was deemed at the time to be the most "international" natural disaster in history, due to its geographic scope and the international nature of its victims and relief agencies, summoning correspondingly immense international interest by the media for years to come (Osborne 2014). The tsunami waves or surges reached the coasts of Thailand, Malaysia, Sri Lanka, India, the Maldives, Myanmar, Malaysia, the Seychelles, and the east coast of Africa. Thus, although the event is known geologically as the Sumatra-Andaman earthquake, it is referred to more commonly as: "the Indian Ocean Tsunami," "the Asian Tsunami," "the Indonesian Tsunami," or "the Boxing Day

Tsunami." With no tsunami warning systems in place in the Indian Ocean, the death toll of the combined tsunami events and associated natural disasters was reported to be over 230,000 people in fourteen countries, putting it among the deadliest natural events of the last hundred years. But the real impact of the earthquake was nonetheless regional and local. Indonesia was the hardest hit of the areas in the Indian Ocean, a fact associated with its proximity to the epicenter of the earthquake and the particular topography of the region that enabled the magnification and inland trajectory of the tsunami waves. The casualties were reported to be as high as 170,000 people, with a quarter of a million people displaced. Most of the fatal victims lost their lives in towns of the western coast of the Aceh province, including Banda Aceh.

Destruction was quick and widespread, but not everything in Banda Aceh was obliterated during the tsunami of 2004. In many areas of the city, and increasing relative to the distance from the coast throughout Aceh Besar, infrastructure still stood. In the affected areas of the city and beyond, mosques stood relatively intact amid debris, a majestic image that repeated itself along the Acehnese territory in the aftermath of the tsunami. As these were often the only structures that remained standing within plain sight, the survivor mosque became a powerful image that provided a clear spiritual message and spiritual relief to the Acehnese. But this message was not sustained by the many aid workers who took great care to explain the different standards of construction that civil engineers used in mosques in comparison to domestic architecture. Other public sites of heritage value survived the devastation at large: the Aceh Museum and adjacent *Rumoh Aceh* (traditional house), government institutions, and places of historical significance. Furthermore, new places of public significance emerged in the years following the catastrophe, a new heritage of sorts. As these emerging sites gained popularity and support through institutional and governmental efforts and public attendance, they proposed an addition to the existing collection of heritage in Aceh. This type of heritage challenged dominant historical narratives that would normally marginalize the inclusion of recent events into heritage assemblages. It also challenged the historical agendas that could manipulate the nature of loss and resilience, as the city and region proposed through this act of valuing heritage a different outlook on the past, present, and future of the city.

Meanwhile, a continually changing heritage assemblage played a key role in the construction of narratives for a new Aceh, as exemplified by the ongoing formalization of the tsunami boats as heritage sites. Lampulo Boat had landed on the second story of a house (Figure 1.1). Precariously reinforced with wooden beams toward improving stability in situ in 2007, it was later stabilized with concrete supports. Makeshift graffiti that dated to the early days of the boat acting as monument read in English, "Don't broken. Tsunami Minggu 26 Des 2004, jam 8:27," a caption that marked the moment in time when the

Figure 1.1 Lampulo Boat, 2010 (source: author).

boat became more than a boat, and also defined the space of the boat as a valuable object to be distinguished, and eventually protected, from the rest of the rubble. By 2008, a copper plaque explained more eloquently in Acehnese, Indonesian, and English: “This fishing boat is located here after being carried by the force of the tsunami on 26 December 2004. The boat’s current position is a very real symbol of the mighty force of the tsunami’s wave. The boat saved 59 people in this incident.” Its inconvenient situation as a subject of preservation efforts was surprising to many. A second tsunami boat, PLTD Apung 1, had been brought in from Kalimantan and used as an electrical generator to supply Aceh. Its journey from Ulee Lheue ended unpredictably inland, approximately 4 km from where it had been moored (Figure 1.2). While in 2007 there was an informal discussion in the reconstruction agency (BRR) about the logistical challenge of returning this ship to sea, by the middle of 2008 the land adjacent to the ship had been secured for the construction of a Tsunami Educational Park. Informal guided tours were then offered on the ship’s decks, from which the entire city could be observed. By 2009, both tsunami boats exhibited an authentic weathered look, characteristic of the unforgiving Sumatran weather. A few houses surrounding PLTD Apung 1 had become gift shops where one

Figure 1.2 PLTD Apung 1, 2010 (source: author)

could buy souvenirs—a scene that was not matched anywhere else in the tourist-starved city. The entrance to this *kampung* now featured a miniature boat replica marking the main access to the site. During the years that spanned this research project, 2006–2011, these sites would show signs of age, but also signs of a dedicated preservation and maintenance that were being applied to them, progressively reinforcing their heritage significance through these simple practices.

On the other hand, older narratives and forms of historical representation that contributed to the presentation of Aceh and the Acehnese in less favorable light were not being necessarily promoted (evidenced, for example, in the lack of attention to the fast decay of the monuments dating to the Japanese occupation after World War II). The emergence and active valuation of a new heritage in Aceh suggested an intention to reorder historical events and imagine a different legacy as it is represented in the future aesthetics and narratives of the Aceh region. This should be seen as a mechanism of selection that is in line with the articulation of cultural resilience of the Acehnese, highlighting the significance of examining construction, destruction, and reconstruction in heritage imaginations here.

Although this complex mobilization goes far beyond the emergence and life histories of specific heritage sites, the approach to heritage taken here is directly inspired by the centrality of new heritage places and narratives in postdisaster Aceh, suggesting that narratives be constructed from new positions, as is the case with the honored boats, from "the deck of the tsunami ship." From this strategic point, the construction and retelling of narratives on Acehnese ground prioritize a post-tsunami chronology that takes into account the centrality of destruction on the built landscape as an opportunity to construct, and on the emerging discourses that give meaning to it.

View from the Deck of the Tsunami Ship

In 1955, J. C. Van Leur set out to reevaluate the presentation of Asian history, accusing his predecessors of viewing the history of Indonesia "from the deck of the ship, the ramparts of the fortress, the high gallery of the trading house" (Van Leur 1955, 261). Two generations of historians have since taken up Van Leur's challenge and debated what it means to view the world of Southeast Asia from a position that acknowledges its own autonomy. This repositioning of history has been central to historiographic assessments of the region since the 1970s, as it seeks to revalorize the region on its own terms and within localized perceptions of the past (Reid and Marr 1979). A tradition of contact and multicultural encounters had made the Southeast Asian archipelago one that was largely perceived as defined by a high traffic of foreign influences, especially from Arabia, Persia, China, and Europe (SarDesai 1989). Because of this, the region had been traditionally understood as the product of processes of "indianization," "islamization," "colonization," and "modernization." These qualifiers obscured indigenous and autochthonous developments in the historic (Andaya 1992) and prehistoric (Heine-Geldern 1972) past, promoting disproportionately the significance of processes of adaptation and innovation relative to these external agents. In brief, the coherence of Aceh was presented as a result of successful vernacularization. But reappropriating Van Leur's metaphor to highlight questions of context and perspective is not entirely original. For example, Enseng Ho discussed a view of this region from "the other boat," looking at a series of contacts between Western empires and Muslim societies through the eyes of a Muslim diaspora, a mobile people with a written history (Ho 2004). Following these footsteps, the approach in this volume appropriates this metaphor to highlight an alternative authorial and authoritative position from the situated perspective of a post-tsunami landscape of destruction and its heritage interlocutors.

In many other parts of the world, traces of an archaeological past have been directly invoked in processes of the legitimization of an indigenous territory and its history. However, the potential for mobilizing an archaeological past in this way in Aceh is weak. Very few archaeological sites have been systematically excavated (McKinnon 1988; Reid 2006; but see Sieh et al. 2015) as there has been little official government support for such work, and local knowledge of such sites where they have been excavated is generally limited. Therefore, scant archaeological evidence from Sumatra points to the identification of archaeological cultures that define this territory in association with wide areas of insular and mainland Southeast Asia and beyond.[2] The closest remains to Aceh of a megalithic culture are the stone village walls and stone seats of the Batak in Toba. However, two Sumatran prehistoric empires hold significance for the history of Aceh. Srivijaya was postulated as the oldest among the Indonesian kingdoms with its center located in the Palembang River valley in south Sumatra from the seventh century onwards (Taylor 2003). The other is Jambi, which Loeb has argued was founded around AD 644 probably as a Buddhist kingdom, but conquered by Srivijaya soon after (Loeb 1972, 7). Negligible archaeological evidence cannot clearly support or contradict the scope of influence of these kingdoms over Aceh (Soekmono 1965), so textual accounts are used to infer their influence and identify the emergence of significant settlements on the northern coast of Sumatra. Early depictions of Aceh therefore rely on textual and cartographic resources that have shaped Aceh historically in its location in space and time but originated elsewhere.

The earliest of these may be a map belonging to the annals of the Liang dynasty that documents the existence of the state of Poli in the north of Sumatra dating to sometime around the sixth century (Loeb 1972, 218), while stories of travel and mobility around the Indian Ocean from the ninth century describe Sumatran ports from the perspective of Hadrami merchants (Ho 2006; SarDesai 1989). Marco Polo visited the port of Samudra by 1292 (Reid 2005a), while in 1323 Ibn Battuta describes the port of Samudra and the region as a sophisticated sultanate with international relations around the Indian Ocean and China (Gibb 1929). Acehnese territory at that time comprised mostly the harbor of *Kuta Radja* (City of Kings) and served as a feudal state to Pasai (Loeb 1972, 218). Portuguese traveler Tomé Pires described the Moor kingdom of *Achin* during 1512–1515 in his work *Summa Oriental* (Cortesão, Tome, and Rodrigues 1944). The kingdom of Aceh also featured significantly in Arabic and Malay texts (Guillot and Kalus 2008; Taylor 2003). Despite these mentions, it has been argued that the early history of Aceh before 1500 remains relatively unknown. The scarcity of written records about the premodern period in northern Sumatra has made it necessary to rely mostly on the context provided by historical accounts and thus made any historical reconstruction in this region arguably weak until the end of the eighteenth century (Soekmono 1965). Moreover, the difficulty of discussing a local identity through historical narratives of the region relates to an academic emphasis on the authority of the written source for the creation and transmission of history. From this perspective, a lack of contribution to, or resistance against, authoritative voices makes it appear as though there is no local production of historical narratives. In this scenario, oral histories, relegated to a position of alternative productions of the past in premodern contexts (Pombejra 2002), construct a version of the past that is deemed as hinging on invalid sources and is therefore considered to be "unauthoritative" and "unhistorical."

While folklore holds a prominent place in the local construction of an Acehnese past, and the earliest cited use of Acehnese language is in the year 1658, the greatest works in that language do not appear until the eighteenth and nineteenth centuries with the epic *hikajats*. *Hikajat Raja-Raja Pasai* is the oldest known text of Acehnese history, telling the epic story of the kings of Pasai that includes the account of conversion of this kingdom to Islam. The legend tells of the lives of two princes, Raja Muhammed and Raja Ahmad, who roamed the north of Sumatra, searching for a place to build a new kingdom. Their eldest grandson, Marah Silu, had a dog named Pasai who caught an ant as large as a cat. On this site Marah Silu built a palace named *Semut Raya* (great ant). This encounter had been predicted by the Prophet Muhammad as one that would give rise to the embrace of Islam in the region and to the founding of the city of Samudera Dar as Salam (Taylor 2003; Leyden 1821, 66). The legend therefore explains the toponym *Samudra*, as distinct from another explanation that derives the word from the Sanscrit for "sea" (Loeb 1972). On Samudra, Marah Silu founded the kingdom of Pasai before 1286, converting it to Islam as he took the title of Malik al-Salih, the first sultan of the region (d. 1297).

Lamri and Pasai are the earliest known Muslim-headed settlements in North Sumatra, both at the entrance of the Strait of Malacca on the Indian Ocean. Arabs are believed to have entered the region between 846 and 950, and with them the Islamic faith. From the ninth century, Hadrami (present-day Yemen) merchants are documented in Sumatran ports (Ho 2006), setting up trading centers in the archipelago by the thirteenth century. The kingdom of Pasai under the reign of Sultan Malik al-Saleh was an Islamic monarchy and became the leading center of Muslim culture in the Indonesian Archipelago, particularly significant in the transmission of Islamic religious knowledge and the production of Malay literature (Feener 2011), looking out toward the Indian Ocean and the China Sea (Taylor 2003). Aceh emerged as one of the several port-states in this area, formed around 1500 on the ruins of ancient Lamri. Despite its commercial and religious prominence Pasai never united the region of northern Sumatra politically, which remained separate Muslim port-states (Reid 2005b). However, the Portuguese arrival in 1509, takeover of Malacca in 1511 and attempt to gain control of Pasai, Perlak, and Pidie in Sumatra drove all anti-Portuguese states to unite. Under Sultan Ali Mughayat Syah

(r. 1514–1530) Aceh developed as an independent kingdom driving away intruders, such as the Bataks of Northern Sumatra, between 1540 and 1630. The acceptance of Islam was a major motivation for these conflicts, while the emphasis on an Islamic network fostered relationships between Aceh and the Ottoman Empire (Reid 2005b), with two distinct periods in which Turkish intervention in Aceh was historically significant. The first took place in connection with Suleiman Pasha's expedition to Gujerat in 1538; the second took place during the peak of Aceh's commercial contact with the Red Sea in the 1560s. It is also in this period that the toponym *Aceh* was extended to Greater Aceh and the provinces in the south.

An era of scholarship about Aceh was ushered in by the colonial presence and activities in the region. Through their agendas, three themes have since figured in the representation of a unique Acehnese heritage—namely, its Islamic identity, its inherently violent disposition, and its intrinsic relationship with the Indian Ocean. All three themes have been critically reevaluated in postcolonial debates, in consideration of the type of sources that construct them (see Stoler 2010), but they have nonetheless persisted as sources for the production of narratives and counter-narratives in the construction of postdisaster heritage in Aceh in the twenty-first century. Chapter 2 in this book will review the types of heritage that allude to these themes, a heritage assemblage that is now accompanied by new heritage constructions of the tsunami era.

While William Marsden, the secretary for the British possessions of the coast of Sumatra (Marsden 1783), presented Sumatra as a relatively coherent unit, Snouck Hurgronje's Dutch-sponsored study of religious elements and the political condition of Aceh resulted in an ethnographically rich volume, *The Acehnese* (Hurgronje 1906), that sheds light on a unique composition of this society. *The Acehnese* defined Aceh as a territory to be conquered by the Dutch at the turn of the twentieth century. It therefore focused on exploiting a rupture in society between *adat* (tradition) preserved by the *uleebelang* (the governing elite) and *hukum* (religion) preserved by the *ulama* (the religious leaders). Further work challenged the idea that Aceh was a society bifurcated into Islamic and customary elements, but rather one divided into *uleebelang, ulama,* peasants, and the sultan and his group—each of which had its own view of the nature of Islam and *adat* (Siegel 1969). Eric Morris further added that the Acehnese leaders, through the conflicts of independence, thought and acted in Acehnese, Islamic, and Indonesian terms, with little awareness at the time of the possibility of conflicts among the three identities that are otherwise presented as contradictory (Morris 1985), highlighting the likelihood of strategic agency in the articulation of identities, rather than an incongruency that needs to be resolved in their characterization. A pivotal factor in the construction of Acehnese identity is the assumed identity of Aceh as *Serambi Mekkah*, the verandah or gateway to Mecca (Riddell 2006). Aceh became the leading regional center of Islamic learning and debate during the seventeenth century (Feener 2011, 9); however, R. Michael Feener points out that Islam "is not itself sufficient to explain the history of the Acehnese culture and society over the centuries" (Feener 2011, 11).

The degree to which Islam is perceived to be at the center of Aceh's distinctiveness is a matter of contention. Historical approaches take a broad perspective, describing a unique relationship of Aceh to other Islamic as well as non-Islamic centers from the sixteenth to the nineteenth century, when interactions with the Indian Ocean and the Malay Peninsula—but not with the Java Sea world of Jakarta—influenced economic, political, and cultural ties (Reid 2005a, 228). For four centuries, the Acehnese therefore thought of themselves as belonging to an independent polity that competed with some comparable states like Siam and Johor. It has been suggested that this debate takes into consideration the peripheral place that Islam has in the region's politics and diplomacy, far from the center of the Middle East (Fealy 2006). But on the other hand, Benedict Anderson argued that the ensuing problem of integrating Acehnese identity and history into the Indonesian state needed to be discussed largely in terms of Islam. He argued that Aceh's inability to accept a vision of Indonesia as their cultural inheritance is central to understanding the failure of the nationalist project in this region and the historical trajectory that followed (Anderson 1999), considering the initial contribution that Aceh made to the revolutionary cause of the republic in terms of manpower and funds without any aspirations for an "independent Aceh." The disenchantment, he argued, came later, and was exacerbated during the New Order, the regime in Indonesia led by President Suharto (1965–1998), when the sentiments of independence in Aceh materialized in long and sustained episodes of conflict and, incidentally, constituted much of the negative portrayal of the Acehnese that was used to justify their oppression.

In association with this disenchantment and the events that surrounded it, many characterizations of Aceh focus on the theme of conflict and postconflict identity politics, with some more recent work on post-tsunami reconstruction recentering debates on the relationship between culture and resilience (Daly, Rahmayati, and Feener 2009). Although varying in nature, conflict has been represented as characteristic, repetitive, and even endemic in Aceh, a mythology that continues to influence the imagination of a post-tsunami Aceh in the way that the disaster has been often articulated within the narrative of punishment for past and recent violence, even where it is represented partially and in conformity with different agendas (Sherlock 2005). As Feener points out, the prolonged conflict with the Dutch left deep scars on both sides. For the

Acehnese, it established the idea of "resistance" as a foundational concept in the formation of their identity. The legacies of conflict during and since the war against the Dutch also fostered popular perceptions and even self-ascriptions of Aceh's history as preeminently one of violence (Feener 2011, 18). In more recent ethnographic work on Aceh-Indonesia relations, Elizabeth Drexler addresses some of the specters of a traumatic past and the role that they play in the continuation of a reimagination of "Acehnese-ness" vis-a-vis agents of violence (Drexler 2008). At a national level, a legacy of political distortion of information around sites and events of state-sponsored conflict has been discussed as an extension of a "tradition of violence" (Fuller Collins 2002), manipulated by elites in order to normalize their own recourse to violence (Zinoman and Peluso 2002). This tradition is constructed in association with strong censorship up until the late 1990s, used to interpret a value for violence in the rise of the regime. It is a violence that hinges on the powerful creation of an *appearance* of a culture of violence through the New Order period (Zurbuchen 2002, 567). Ann Laura Stoler has pointed out that a cycle of continuous rehearsals of the violence associated with the infamous 1965 communist killings in Indonesia overshadows and silences other pasts that are only beginning to be reimagined (Stoler 2002). In addition, Mary Zurbuchen convincingly argued that the violence that occurred between the 1960s and the 1990s is embedded in an even deeper set of colonial histories and representations (Zurbuchen 2005), but contemporary "sites of memory" are able to contest productively the authority of written narratives. In Aceh, the episodes of conflict against the Indonesian state had not yet resulted in physical markers in the landscape or a conciliatory heritage of reparations. Instead, the legacy of conflict, such as the location of graves of the deceased in an illegitimate combat, was circulated in rumors and oral history (see Martinkus 2004). The fall of Suharto in 1998 presented an opportunity to discuss the politics of commemoration in Indonesia (Dittmer 2002), in consideration of the state's role in the construction of official history (Cribb 2002) and through the conjuring of negatively associated subversion (Zurbuchen 2002). Hence, in Aceh, representations of a collective memory of the tsunami are inexorably tied to representations of a censored experience of the separatist conflict that unsettle the process of remembering (cf. Huyssen 2003).

Finally, the nature of the 2004 disaster revived a discursive significance of Aceh's link to the sea. One constant that is summoned in the history of Aceh is a relationship with the Indian Ocean that forged trans-insular links and cultural contacts. In fact, Indonesians claim an ocean relationship as a unique identity nationwide, calling their land *tanah air* (land of water). The relationship to the ocean helps to construct a unifying identity for a nation built on an archipelago of over thirteen thousand islands, featuring widely varying environments and ecosystems as well as significant cultural distinctions across its territories. But the construction of Indonesia as Asia's "most maritime region," Anthony Reid argues, is a product of a colonial emphasis on the coastal areas that excluded inland livelihoods (Reid 2005a, 48). In Sumatra, Reid argues that this activity is evidenced by impressive upland agriculture prior to the nineteenth century and the presence of pre-Islamic monuments in highland plateaus. In an inverse situation, evidence of past experiences of ocean-borne natural disasters is not featured in the natural and oral history of Indonesia or many of its regions. Considering an Indonesian geography that can be said to be sculpted by the forces of nature and amid the so-called Pacific Ring of Fire, it is perhaps surprising that disasters are not featured more prominently in regional identities within Indonesia. Overviews of the region do not delve into such environmental conditions and concerns (i.e., Frederick and Worden 1993; Hardjono 1971; Peacock 1973), an absence that suggests that this type of event is not significant in shaping regional geopolitics. However, in Sumatra, geo-modeling suggests that the earthquakes of 1833 and 1861 were likely to have generated tsunamis (Dominey-Howes, Cummins, and Burbidge 2007). Additionally, the study of sediment records on the west coast of Aceh (Monecke et al. 2008) reveals two tsunami deposits in AD 780–990 and AD 1290–1400, and possibly a third in AD 1907. Recent work has begun to identify a significant number of other such seismic events in the past, and their influence on prehistoric occupation patterns (see Sieh et al. 2015). Nontheless, tsunamis are a rare occurrence and difficult to correctly identify in historical records. Prior to the catastrophe in Aceh, there appeared to be little notion of this phenomenon, with the exception of an 1893 incident that remained in the collective memory of a specific region of Aceh. On the nearby island of Simeulue, closer to the epicenter of the same 2004 earthquake, an extremely low number of casualties[3] was associated with the effectiveness of *smong* stories, transmitted through oral history, since a 1907 earthquake struck the island (McAdoo et al. 2006). The absence of a comparable mechanism in the heritage of mainland Aceh puzzled researchers and practitioners alike, who questioned the effectiveness of their work toward building resilience for future tsunamigenic events. The observed failure suggested that new cultural heritage has a key role to play in the establishment of a disaster-sensitive culture and education.

Heritage as History or Heritage as Witness

Against this historical background, what heritages might be constructed in postdisaster contexts such as the one that faced Aceh following the 2004 tsunami? Dominant heritage constructs lean excessively on a historical construction of significance, a *longue durée* approach in which the recent and the present cannot acquire heritage value, and changes can be understood only as the product of long-term processes. Complex disasters, such as the 2004 tsunami, are unexpected and sudden, often unrelated to other historical or natural processes in operation. An alternative framework is therefore needed in order to identify, capture, and debate the construction of a safeguarding agenda for these emerging heritage constructs in postdisaster settings characterized by sudden change. In archaeology, a recognition that change may have multiple temporal and spatial scales rather than align coherently with arbitrarily imposed measures of significance has focused attention on the relationship between people and their environments according to different levels of perception (Torrence and Grattan 2002b). Torrence and Grattan (2002a, 9) have proposed taking into consideration the experience of a hazard from multiple temporal and cultural perspectives, including the role of oral history and mythology in shaping this experience. The "heritage of disaster" I describe here puts destruction and its associated narratives at the center of processes of heritage making through the nature of the materiality it produces or alters, but also through the roles and performances that this type of heritage engages. This is a type of construct that could equally comply with, or resist, planned and intentional forms of commemoration, emerging—as the tsunami boats did—organically, and existing at the intersection of traditional and contemporary mastery of the heritage ideal. In this way, destruction might be preserved, made visible, and appropriated into constructive roles (see Holtorf 2015).

In a heritage of disaster, heritage construction and traumatic memory intersect more evidently as social phenomena (Halbwachs and Coser 1992). In this particular context, narratives associated with a place become integral to the formation and inscription of memories in contemporary society (Severi 2001) but also become transmitted trans-generationally to descendants who are able to reiterate these memories (Ebron 2007). There are two roles that heritage construction takes in postdestruction contexts: following Ricoeur's *récit* (1990), two performative aspects are involved in the construction of social memory—the work of narration, and the elaboration of ritual practices that imbue objects and places with meaning. The creation and preservation of heritage places should be seen as one of these ritual practices and the vessel through which narration is transmitted: the repeated, habitual actions that take place in the transmission of cultural knowledge (Connerton 1989). The second role that a heritage of disaster takes is that of a witness, since trauma (like heritage) is collectively and individually transmitted as a socially mediated category informed by imagination (Alexander et al. 2004), challenging the notion that events are inherently traumatic. It highlights the importance of cultural and social structures and contexts and is supported by research on "witness studies" (Douglass and Vogler 2003; Whitehead 2007). In Indonesia, witness studies has focused on the 1965 killings in Bali as the center for an emerging discourse of trauma as mentioned earlier. This is a discourse that is woven into the histories of Southeast Asian nations, as postcolonial states come to terms with internment camps (Coté 2009), torture centers (Long and Reeves 2009) and sites of imprisonment (Leach 2009)—sites whose preservation and interpretation demand a clear identification of the perpetrator, becoming entangled in the politicization and historicization of the landscape. Because of this entanglement, witness studies enable a discussion of the morality of place, raising concerns about the hierarchization of different types of historical legitimacy, as factual and subjective accounts must be integrated (Young 2003).

It could be argued that Southeast Asia is not a viable context for the exploration of heritage as witness, as heritage is largely regulated by state agendas and the foreign consumption of traditions through tourism (Hitchcock and King 2003). Denis Byrne's research on the history and politics of heritage management in Southeast Asia offers a compelling explanation for considering heritage an unfit construct to act as witness, pointing out the elusive visibility of heritage in archaeological remains (Byrne 1997) and the inability of loss to be converted into tangible heritage constructs (Byrne 2007; see also 2014). Byrne's work suggests that heritage discourses and constructs have been complicit in burying the past through an encrypted terrain, taking part of the conceptual underground instead. Notwithstanding these caveats, the heritage of disaster in Aceh highlights the potential role of cultural heritage as an eyewitness device. Both material culture and narratives associated with cultural heritage places can constitute mnemonic devices that transmit, validate, and resist vernacular histories of trauma over time (cf. Zimmerman 2007). Material culture, in this sense, acts like any other form of media: it collects, reshapes, and transmits events in an ongoing process of explaining the past (Morris-Suzuki 2005). Some have rejected the idea that architecture is capable of embodying memories (Forty and Kuchler 1999), but the proliferation of sites of memory and "negative heritage" around the world suggests, on the contrary, that material culture can play a key role in enabling an experience of memory and trauma. It fulfills this role through its living performance and active preservation, where the practices associated with its ongoing maintenance reinforce a significance bestowed on them and a relevant place in society. Due to a continual process of construction in which heritage is engaged, traumascapes that are constructed alongside heritage values and discursively attached to heritage assemblages have the ability to be sustained through time, breaking the monotony and artificial coherence of a historical approach to heritage assemblages.

As a city that is continually being constructed and deconstructed in a careful interplay of influencial exogenous sources and contemporary localized processes, Banda Aceh presents itself as an opportunity to highlight the value of a postdisaster landscape "in progress." While the heritage that emerges in this city resists any single trajectory and narrative, what remains notable is the process of assimilation and countering of destruction that are being expressed in creative ways which, I argue, enable the reconceptualization of Aceh as a coherent space and identity. This is an ingenious appropriation of "negative" heritage that moves away from using the identifier of catastrophe to victimize identities; instead, it mobilizes destruction as a tool for carving a space in which the conversation about the future of Aceh can carry on less burdened by the past. This requires, however, a validation and protection of a contemporary and ongoing production of heritage. The following chapter describes in more detail the interplay between heritage representations and different historical trajectories that are documented for Aceh, considering the marginalized position in which a heritage of the contemporary is located.

2 Heritage Narratives in the Tsunami City

An examination of the construction of postdisaster heritage landscapes challenges the predominant idea that history forms the backbone of constructions of heritage, complicating the way that heritage and history coconstruct places and identities. The city of Banda Aceh through each historical and contemporary iteration has produced different types of cities, up until the proclamation—perhaps temporary—of the city as a tsunami city, mobilizing and remixing history in strategic ways. This chapter centers on the role that heritage possesses as a subject that informs these constructions, used as an assemblage that actions different degrees and directions of destruction for specific purposes and in accordance with the roles and visibility that are given to different historical moments in Aceh. Retelling the history of the making of Banda Aceh in historical terms, this chapter utilizes heritage as a vehicle for the expression of a unique relationship between the Acehnese and their historic environment that has enjoyed different degrees of autonomy and visibility, materialized through prehistorical, historical, and posthistorical stages of heritage construction.

From *Serambbi Mekkah* to Tsunami City

Banda Aceh was not always a "tsunami city." There are plenty of historical narratives that strengthen a sense of distinctiveness in Aceh under different denominations and continuously recreate its sense of uniqueness, shaping the construction of its heritage assemblage (Harrison 2013; MacDonald 2009). However, in attempting to define and describe this place at different times, I aim to make the distinction between heritage constructions informed by archaeological and historical knowledge and the heritage assemblages that are constructed through more subtle narratives embodied and transmitted in monuments, toponyms, and practices. This distinction considers that constructions of the past, present, and future should be further problematized in the heritage of Aceh. I conceptualize the interpretation and presentation of different pasts through a distinction between a "preheritage history," a "heritage history," and an "anti-heritage history," distinctions that I make not necessarily in association with different chronological periods but as a result of different contexts in the construction of heritage that result in more or fewer remnants and heritage constructs from that period in the life-history of Aceh. These analytical categories also highlight the active role that heritage constructs have in bringing coherence and transferability to historical moments in accordance with sociopolitical contexts that give different histories their degree of legitimacy.

A preheritage historical period could be said to predate the construction of tangible cultural heritage, although it may be officialized as heritage *post facto* to inform and strengthen historically informed heritage. This cannot be simply referred to as the better known concept of "archaeological heritage" because archaeological heritage itself is built on existing and recognized remains, however scant or under discussion, which have already been processed and constructed as a type of heritage. What I suggest in this phase in the history of Aceh is to distinguish a chapter in the history that predates its documentation, as has already been discussed, and which has therefore failed to leave a mark on the built landscape and tangible heritage remains—it exists only in textual sources and debates, in unexcavated archaeological remains, and in oral histories, remnants that are immaterial regardless of their consented legitimacy. However, while this type of history may not be directly associated with existing tangible heritage products on the ground, it may play a key role in the imaginations of place that are made to be coherent with a broader historical narrative, as I will describe in the next section.

A "historical heritage," by contrast, is constructed in close association with an assemblage of tangible heritage places, where history awards heritage value to material culture, and material culture legitimizes history as fact. It is an integral part of the construction of an archaeology of memory (Lowenthal 1985): a trilogy of memory, history, and relics that constructs a vernacular geography through coherent narratives. Such a framework highlights the significance of a practice of bringing together pieces of the past that are at once timeless and discontinuous. A past with a heritage is a powerful past, as it continues to be used in negotiations for self-representation. Because of this, "historical heritage" is not all-inclusive, and should be seen as a carefully constructed assemblage that is subject to further alterations.

Finally, there is a type of assemblage that can be referred to as an "anti-heritage history." This type of heritage addresses two types of absences in the formulaic interaction of heritage and history. On one hand, an anti-heritage refers to an erased history that has been intentionally or accidentally marginalized in a heritage assemblage. As Denis Byrne has argued in his volume *Surface Collection*, heritage landscapes may conceal an encrypted terrain (Byrne 2007, xii), suggesting that heritage is an inadequate construct through which to identify an experienced past. On the other hand, an anti-heritage is also the excluded byproduct of the work of a dominant heritage construction that marginalizes any recent history from taking part in heritage assemblages for preference of a historical value in the construction of heritage. This preference is reinforced particularly by a dominant discourse that informs in practice myriad legal and disciplinary frameworks that determine which heritage can be legitimized in chronological terms. The dominant framework is unwilling, and therefore unable, to bestow heritage value to the present. Therefore, a heritage in which the emphasis is on the contemporary holds an uneasy place in heritage studies and practices (cf. Harrison 2011). It is a heritage that works more creatively at the boundary between past and present (Little and Zimmerman 2010). As Barbara Little and Larry Zimmerman (2010, 142) have convincingly argued in defense of agendas of the contemporary, "time is not the important variable, but the uses and distribution of cultural materials are". Hence, a heritage of the now has the potential to resist, rewrite, or enhance historically established trajectories.

Heritage History

One of the earliest officially recognized tangible heritage objects in what is the current territory of Aceh pays homage to the significance of early trade with China in Sumatra, a bronze bell "Cakradonia" that adorns today the front of the Museum Negeri Aceh and the Rumoh Aceh complex. A gift from the ruler of Zhenh He's kingdom in China in 1414 to Samudera Pase, it was brought to Aceh after the fall of Pase in 1526. Ironically, despite a relationship that likely dates back as far as 1000, the Chinese are not well represented in the history and heritage of Indonesia as a whole. They have been marginalized on the grounds of being a nonindigenous ethnic group, and have even been the subject of targeted aggression (Purdey 2006). A row of *Ruko China* shop houses in the Peunayong quarter downtown is evidence of Chinese presence in Banda Aceh, which is historically known (Reid 2005a). But the heritage landscape that is associated with the early history of the territory that eventually became Aceh (and which is articulated as part of the genesis and heritage of Aceh today) is dominated by the history of Islamization. Muslim tombstones imported from Cambay in Gujerat date back to the earliest periods of Islamic presence in Sumatra (Heine-Geldern 1972, 327). Through these monuments, the movement of a Hadrami diaspora across the Indian Ocean territory is mapped in Sumatra, and in particular, Aceh (Ho 2006), unfolding the early history of the sultanate and the region. Following the rise of Aceh described in the previous chapter, the sultanate's prosperity peaked at the end of the sixteenth century (Hadi 2004), a golden era under the reign of Sultan Iskandar Muda (r. 1607–1636), who is said to have constructed *Taman Sari*, a royal complex whose gates, "Pintho Khop," can be visited today. His remains rest in the Royal Mausoleum at Kampung Pade. His successor, Sultan Iskandar Thani (r. 1636–1641), built in this complex an artificial mountain, the Gunongan, for his wife Putro Phang, embodying the significance of the role of *tasawwuf* (spiritual development) in expressions of religious and political culture of the sultanate (Feener 2011, 8). Today, Iskandar Thani's grave is located in the Kandang XII complex in Keraton Village alongside twelve other mausoleums.

In this period, new forms of Islamicate art and culture could also be observed, demonstrating the influence of the Mughal court that took root in Aceh (Feener 2011). This process of transformation also attracted Islamic scholars from the Middle East and South Asia, and inspired the establishment of local prominent scholars. The *ulama* Hamza Fansuri (d. 1590) is often celebrated as the "first teacher" in Sumatran Sufi literature and remains a significant figure in the landscape of post-tsunami Aceh. His grave was allegedly found during post-tsunami landscape surveys: a long grave in the hills at the northernmost tip of Aceh was previously within the dangerous terrain of separatist forces. Although its authenticity had not been confirmed, the site is now approached by a paved trail, leading to a clean, draped, and enclosed grave that is accompanied by a new *mushollah*. Fansuri's writings were opposed by Nur al-Din al-Raniri (d. 1658), a Gujerati Muslim who established himself at the Acehnese court in 1637 and initiated a campaign of religious reform toward a more orthodox doctrine (Feener 2011, 10). His work sought to redefine Aceh's Malay Muslim tradition in the fields of jurisprudence, mysticism, theology, literature, and history. The decentralization of the sultanate later had an effect on cultural production in Aceh. While the Malay-language Islamicate culture had found an important center in the centralized sultanate, by the eighteenth and nineteenth centuries, however, this literature is observed to have fewer religious elements (Drewes 1979, 9).

The European influence in the material culture of Aceh began in the mid-1500s, when Aceh became a target for the Portuguese, and later the Dutch, the English, and the French, who sought the abundant pepper of the region. The Dutch East India Company (VOC) was formed in 1602, with initial interests over Malaysia. Despite its power and wealth, Aceh was not able to retain full autonomy. Sultan Iskandar Thani formed an alliance with the VOC to drive the Portuguese out of Malacca. By 1650, Europeans had gained control of the vital ports and products that had previously linked the region to the expanding world economy (Reid 1993). The Aceh sultanate was forced to sign an agreement with the VOC in 1659, giving the company a trade monopoly. The waning of the kingdom of Aceh occurred during the reign of four female sultans who ruled in rapid succession during 1641–1699, amid diplomatic tensions between VOC officials and the transition to the rule of the sultanas (Khan 2011). The abdication of the last sultana in 1699 gave rise to a period of dynastic conflict. A short-lived Arab dynasty achieved relative stability under Jamal al-Alam Badr al-Munir (r. 1703–1726), but this was overturned by the Bugis leader Maharadja Lela Maulajo, who assumed the throne under the name of Sultan Ala' al-Din Ahmad Shah (r. 1727–1735), a struggle for power preserved in the *Hikayat Potjut Muhamat* (Drewes 1979).

The Dutch wars in Aceh that followed this period (1873–1942) are the most historically documented period in the region's history (Feener 2011, 17). The replacement of Sir Thomas Stamford Raffles's Anglo-Aceh treaty in 1819 with the 1824 Anglo-Dutch Treaty of London reformulated the Dutch and British influence in the region after the Napoleonic Wars. Under this treaty Sumatra remained a Dutch possession, and despite initially preserving the autonomy of Aceh, the Dutch ultimately advanced on north Sumatran territory under a new Anglo-Dutch "Sumatra Treaty" in 1871. The Dutch sieged Aceh on 1873, but the citadel was abandoned in early 1874 in the midst of a cholera outbreak. The Dutch ruled Aceh for the following twenty years, but the sultanate held out for thirty years until Sultan Muhammad Daud surrendered in 1903.

The colonial presence of the Dutch on the urban landscape of Aceh was strategic, characterized by the construction of infrastructure on established sites in order to inherit excellent locations and trade reputations. At first, fortresses and buildings were erected in the native style, using wood, bamboo, and *nipa* palm thatching, but later more durable and imposing materials and buildings were adopted. The new physical appearance of the European-controlled cities came to symbolize the strong and seemingly permanent presence of the Europeans in the region (Andaya 1992), evidenced in the colonial heritage structures that remain today. The governor's house, Pendopo Gubernir, was built in 1880 on the site of the palace of the Acehnese sultan. Other landmarks of Dutch heritage include the Dutch communications building, a Dutch water tower, a Dutch water storage building, the former railway station, and a series of Dutch residences that remain in Blang Padang and Keraton, as well as a Catholic church. Even the Mesjid Besar Baiturrahman, originally built in the early 1600s at the center of Banda Aceh and destroyed in 1873 as a casualty of the Dutch invasion, displays the mark of Dutch presence in the area. It was reconstructed by the Dutch government by 1883, allegedly in an attempt to end hostilities and accomplish their political aims (Bouwsema-Raap 2009). At this time the mosque was designed by an Italian architect in the Moghul Revival style of Europe that was first introduced by the Dutch and the British, and therefore was not welcome (ibid., xvii). Subsequent renovations in 1935, 1958, and 1982 added elements to the original reconstruction, but the colonial style mosque was appropriated and reproduced throughout the province. The original vernacular mosque design in Aceh was a tiered-roof design, as can still be seen in the mosque of Indrapuri.

Anthony Reid explains that Aceh was, by far, the biggest challenge to Dutch arms and diplomacy. Their resistance at the turn of the century resulted in ten thousand Dutch casualties, and five times as many Acehnese (Baehr 2008). The cemetery Kerkhof Peutjoet was established in 1880 as the resting place of more than 2,200 soldiers (mostly Dutch, but also Javanese, Menadonese, and Ambonese). The Dutch policy was thus known for its force and aggression, a position that has been attributed to being perpetually under siege by the Acehnese. The Dutch faced episodic resistance until their surrender to the Japanese in 1942 (Reid 1979). Early Acehnese resistance was led by Teungku Shaykh Saman di Tiro (1836–1891), a militant leader who rallied under a religious banner, relying on the popularization of *Hikayat Perang Sabil*, an Acehnese epic that focuses on martyrdom. The Dutch retaliated systematically, and in the period between 1898 and 1914, twenty-four thousand Acehnese were reportedly killed in conflict. While under Dutch occupation, Aceh appeared to resemble and integrate into the rest of the Indonesian colony, connected by the new trans-Sumatran highway in 1938 and coordinated under the first pan-Sumatran administration with a governor seated in Medan. Aceh was instrumental, nonetheless, in rallying anti-Dutch sentiment and raising nationalist awareness during this period and up until the arrival of the Japanese during World War II.

The occupation of Indonesia by the Japanese Empire during World War II (1942–1945) is remembered as the most oppressive colonial regime in the young nation's history. On the other hand, this invasion challenged Dutch colonial rule, making it possible for Indonesian independence to be conceived; in this sense, the Japanese brought about an unprecedented political awakening. This is perhaps why this negative period in the history of Aceh is nonetheless memorialized to some extent. Notably, in the area of Kecamatan Darusssalam, towards the northeast, a Japanese landing monument became part of the Aceh Heritage Trail (Arif 2008). The monument's rhetoric was reminiscent of World War II propaganda, and the monument remained in a deplorable condition, a reminder that monument construction is vulnerable to fictitious significance and therefore may suffer a short life span. In a different part of Aceh Besar, Japanese concrete pillboxes were still standing, but rapidly decaying along the coast under the waves of Ujung Batee beach. *Proklamasi Kemerdekaan Indonesia* (independence) was announced when the Japanese issued an unconditional surrender to the Allies in August 1945, commemorated as well in a monument around Taman Sari in Banda Aceh. At this crossroad, the Dutch attempted to reestablish their rule. They managed to reoccupy most of the territory of Indonesia in a move that led to an Indonesian War of Independence that ensued for over four years. This was a period marked by political upheavals, including guerilla struggles and international diplomatic interventions. Aceh's role as the last remaining independent state during the attempted Dutch takeover period would come to play a key part in its rhetoric of uniqueness and the struggles for autonomy toward the Indonesian nation-building project. It was also instrumental for justifying a rupture between official national histories and heritages and the expression of local distinction and autonomy. The conspicuous RI-001 Monument located at the southwestern end of the Blang Padang square in downtown Banda Aceh represents this struggle. *Seulawah* (Gold Mountain, in Acehnese) is a DC-3 airplane that Aceh "contributed" to the independence efforts of the Indonesian state (Figure 2.1). The plaque on the monument summons the "sincere appreciation from the Indonesian national army air force to the people of Aceh in order to uphold and defend the independence of a united Indonesia, based on Pancasila and the 1945 basic law during 1948–1950." The monument

Figure 2.1 Seulawah, Banda Aceh (source: author)

and plaque, dated July 29, 1984, stand ambiguously as a reminder of Aceh's commitment to the national project of Indonesia, despite the separatist conflict that ensued (Drexler 2008). As such, it can be placed adequately at the transition point between official and unofficial heritage constructs that exist in Aceh, but that have been made to partake in different officializations of history. A crisis of heritage construction and representation unfolds in Aceh from the attempt to have a state-sponsored official heritage and a locally constructed heritage that coexist within the same heritage narrative.

Anti-heritage History

A monolith in Pulau Weh (Sabang), called *Kilometer Nol* ("Km 0"), marks the northwestern limit of the territories of the archipelagic republic. It was inaugurated in 1997 by then vice president Try Sutrisno, but by the end of 2002, journalist John Martinkus observed it to be deserted and derelict (Martinkus 2004, 3), a sign of the significance that Indonesia's Kilometer 0 assumed in the separatist efforts of Aceh when the conflict intensified. As the Indonesian state formalized its hold on Aceh and Aceh's separatist identity strengthened, the relationship between local history and heritage became further complicated, raising the question of the ability of local and national heritage assemblages to represent an indigenous memory and experience of history in Aceh.

This tension is echoed in the crisis that faced early expressions of an Acehnese autonomous identity against *Pancasila*, the official philosophical foundation of the Republic of Indonesia and a state philosophy that competes with Islam and its commitment to Allah, Muhammad, Islamic law, and a place for Indonesia—within the global Muslim community (Taylor 2003, 2). In this crisis, the legitimization of different cultural legacies from the past plays a key role. The kingdom through which nationalist historians articulated the construction of Indonesia's past was far from Aceh, but rather in the thirteenth-century Hindu-Buddhist kingdom of Majapahit in modern Surabaya, East Java. Twentieth-century nationalists, such as Yamin (1962), found in the Dutch interpretation of the epic poem from this era, the *Desawarnana*, evidence for an archipelago-wide political and cultural unity that existed for centuries prior to colonial rule. On this interpretation Yamin built the dreams of a successor state to Majapahit, whose legacy remains etched in Indonesian arts and ceremonies, as well as in texts in Sanskrit and Old Javanese (Taylor 2003). Incidentally, this is the type of cultural heritage that is chosen to represent the Indonesian state on the World Heritage List: the monumental remains of Prambanan, a ninth-century Hindu temple, and Borobudur, an eighth-century Buddhist complex, both located in Java. It has also been pointed out that the heritage preserved and presented for tourism and local consumption in Indonesia, in general, is detached from histories experienced by the citizens of the largest Muslim country in the world (Taylor 2003, 28); however, Taylor also argues that in Indonesia an Islamic past is not experienced through material culture, but through a heritage of stories, legends, and traditions (Taylor 2003). The national motto of the republic, *bhinneka tunggal ika* (unity in diversity, from the Old Javanese), expresses the Javanese belief in the essential unity of all things, while at the same time positing it as a nation-building project that reconciles traditional and modern elements (Darmaputera 1988) and, above all, harmonious integration. But this reconciliation was not immune to regional sentiments of distinction. In Aceh, the neighboring Islamic monarchy of Pasai—contemporary to Majapahit—was the preferred kingdom through which the new nation could be imagined. A kingdom of Javanese ruling non-Javanese was not a vision that appealed to all Indonesians, but several strategies were employed to appeal to the masses. Malay-language narratives, *hikajat* and *sejarah* (histories), played key roles in the blending of Islamic conversion and nationalism, providing a link between the sultanate and Majapahit (Taylor 2003, 104). Genealogies were constructed to show the descent of the ruling class from local gods, rewritten to fit the gods into an Islamic scheme, a scheme that leads to Muhammad's revelation (ibid., 107). Narratives such as *Hikayat Radja-Radja Pase* were key in the establishment of a realm of "fact" between *hikajat* and history (Siegel 1979).

As early as 1951, as the Aceh province was incorporated into the republic's province of North Sumatra together with the ethnically distinct and Christian Bataks, some Acehnese felt that their needs were not being met by the creation of a unified Republic of Indonesia. The omission of *syariah* (Islamic law) in the 1945 national constitution in Indonesia (and subsequent bids to implement it in 1959, 1967, 2001, and 2002) highlights Islamic identification as a discordant element in national affairs of the young nation (Fealy 2006). Therefore, Aceh sought autonomy in matters pertaining to traditional values, such as religion, culture, and education. After independence, anti-Jakarta sentiment was manifested with the Darul Islam rebellion under the leadership of Daud Beureu'eh, who declared Aceh as part of the *Negara Islam Indonesia* (Islamic State of Indonesia) in 1953. The movement hoped for a greater degree of autonomy within an Islamic Republic of Indonesia—specifically, an Islamic federal state, but not necessarily a separate nation. A new monolith now marks the approximate place where an end to resistance was attempted: "On this site, in 1957, a peace agreement was reached between Indonesian Darul Islam and the government of Indonesia," along a road that appeared to have been named relatively recently as *Jalan Ikrar* (Pledge Road). The site itself had no tangible remains to commemorate; it was a material resuscitation of a historical turning point, and not a very optimistic one. The Darul Islam movement was politically undermined when Aceh was granted the status of *daerah istimewa* (special region) in 1959, allowing it local autonomy over religion, customary law, and education. Another major point of dissatisfaction with the national project was the use and distribution of Acehnese natural resources, managed by the central government and foreign partners. The discovery of natural gas in the Acehnese area of Lhokseumawe in the 1970s affected dynamics between Aceh and the Indonesian government under Suharto's New Order (1966–1998).

A new era in the struggles for independence was led by Mohammad Hasan Tiro, who founded the *Gerakan Aceh Merdeka* (Free Aceh Movement, GAM). GAM proclaimed the independence of Aceh clandestinely on December 4, 1976, asking for a different sovereign outcome for the region, and initiated a campaign of armed resistance in the face of an intense Indonesian counterinsurgency. This resulted in the escape and exile of many of its leaders. In 1989 a reemergence of GAM resulted in Aceh being declared a *Daerah Operasi Militer*, DOM (military operations zone), an acronym synonymous with violence and unrestrained and unaccountable military actions that continued until the end of Suharto's New Order in 1998. After Suharto's resignation in May 1998, increased support for the Free Aceh Movement resurfaced in Banda Aceh. A new democratically elected president, Abdurrahman Wahid, called for reconciliation the following year, apologizing for previous abuses committed. The independence of East Timor in 1999 and revelation by the free press about atrocities committed by Jakarta in Aceh created a climate of opportunity for raising the question of Acehnese independence through referendum. Peaceful marches were reported in Aceh and Jakarta, while in October 1999, Wahid established the Independent Commission to Investigate Violence in Aceh, with some officers and soldiers convicted for killing civilians (Sherlock 2005). Wahid's government also undertook two initiatives of importance in 2000 and 2001: a ceasefire, which was referred to as a "humanitarian pause," in June 2000; and the passing of a law for "special autonomy" for Aceh in 2001 that would authorize certain elements of *syariah* law in local courts, a direct election of the province's governor and district heads by 2004, and an increase in oil and gas revenues for the province. A Cessations of Hostilities Agreement (CoHA) brought peace to Aceh at the end of 2002, but by the following April, militias of undetermined denomination began attacks on international teams, causing President Megawati Sukarnoputri's government to seek a military solution. Progress in negotiations faltered in March 2003, bringing an end to the ceasefire on May 19, 2003, when Indonesian security forces launched an all-out military offensive. Sukarnoputri then issued a decree declaring a state of emergency in the province for six months, authorizing the *Tentara Nasional Indonesia* (Indonesian National Armed Forces, TNI) to take responsibility for security and launching an operation involving up to fifty thousand military troops and police against a few thousand GAM guerrillas. Critics of Megawati, the daughter of Indonesia's first president, Sukarno, see this as a continuation of Suharto's policy during the New Order regime of 1965–1998. A declaration of martial law by presidential decree signed by President Megawati banned all foreigners from Aceh in 2003 (Martinkus 2004), relegating Aceh to a period of relative invisibility. Without a doubt, this period affected the ability of Acehnese to construct or validate locally emergent forms of commemoration and narrative with heritage potential.

In this state of emergency, the tsunami catastrophe of late 2004 struck an Aceh that had endured approximately thirty years of heightened conflict. Central government and foreign aid took charge of the region, as representatives of the government of Indonesia and GAM met for the first time since 2003 to address Aceh's needs. For the first time in the history of this conflict, the Aceh situation was being witnessed by an unprecedented influx of foreign observers.[4] GAM welcomed a foreign presence, while the TNI continued and even escalated the campaign against GAM in the days immediately following the tsunami (Kingsbury 2006, 1). A memorandum of understanding (MoU) between the government of the Republic of Indonesia and GAM was finally signed in Helsinki, on August 15, 2005, during the Helsinki Peace Process. In this agreement, the government of Indonesia made concessions in matters of local politics and security agreements. In accordance with this process, the European Union dispatched three hundred monitors to the region to oversee the implementation of the MoU until December 2006, supported by the Association of Southeast Asian Nations (ASEAN). The MoU also requested that former GAM fighters be disarmed and TNI troop levels reduced within the territories of Aceh. This agreement was reiterated at the Indonesian parliament through Law 11/2006 (Aspinall 2008). As determined by the conditions of the Helsinki agreement, democratic elections were held in Aceh in December 2006, for the first time in three decades and allegedly becoming the most participatory Indonesian election ever (Miller 2009). A former GAM rebel and elephant veterinarian by profession, Irwandi Yusuf, was elected as Aceh's new governor. Escaping jail during the tsunami, Yusuf fled to Sweden, where much of the GAM political leadership resided in exile. Only after the signing of the historic Helsinki peace agreement did Yusuf return to Aceh. The post-tsunami political context opened up two possibilities for the construction of heritage in Aceh. On the one hand, the terms of the MoU agreement allowed the Acehenese to express their unique identity in ways that were previously restricted, due to a tradition of censorship and oppression. On the other hand, the opening up of a previously isolated Aceh to the presence and influence of a global community made the region vulnerable to externally imposed heritage discourses and agendas that could aim to bring coherence and homogeneity to Acehnese heritage in consideration of a perceived Indonesian heritage construct and approaches. Therefore, as Aceh was free to express uniqueness through autonomous forms of cultural heritage, these expressions could nonetheless resist full compliance with external norms and standards.

Disaster Legacies

The Agency of the Rehabilitation and Reconstruction for Aceh and Nias (BRR) reported in 2005 a summary of infrastructure damage caused by the earthquake and tsunami of late 2004: 190,000 houses, 14 sea ports, 120 bridges, 3,000 km of roads, 1,052 government buildings, 2,000 school buildings, 114 health centers, 20,000 fish ponds, and 60,000 ha of agricultural lands.[5] BRR had been established under a state of emergency by the Indonesian Government Regulation No. 2/2005 on April 16, 2005 (later Law No. 10/2005). BRR was given the responsibility of redeveloping Aceh and Nias, managing projects funded by the government's annual budget over a four-year period, and coordinating projects funded by donors and foreign NGOs. BRR folded its operations as planned in April 2009, and many NGOs followed suit, leaving behind a largely reconstructed city and its surrounding areas. Henceforth, the landscape of Aceh boasted the hopes of recovery while navigating the traumatic recent history of its making. This traumascape was aptly constructed on key tsunami-era sites of memory. It could be said that the biggest legacy of the catastrophe for the heritage landscape of Aceh was intangible. As described in a BRR report of December 2005, the tsunami was a "virtuous circle"[6]: it created a chance for peace that the reconstruction effort strengthened by bringing the entire community together to plan for their future. Significantly, in this case, the "sense of community" included other nations, which were considered integral to the construction of this new phase in the history of Aceh. A community of international cooperation was in fact commemorated on the Thank the Nations Plaza, across from the Aceh Tsunami Museum. Before the tsunami, this central space was known as Blang Padang, reportedly owned and used by the TNI for displaying military exercises and displaying the *Seulawah* airplane monument. After the tsunami, this area was reappropriated for the display of tsunami commemorabilia that now adorned the whereabouts of *Seulawah*: a walking path to guide visitors through a perimeter of monumental "thank you" notes, concrete *tugu* (monuments) in the shape of a boat that emerges from the ground, each one representing a nation that contributed to the recovery of Aceh. Each plaque includes a flag, the name of the nation, and an inscription that reads "thank you" and "peace" in the corresponding language (Figure 2.2). This collective effort of post-traumatic solidarity was summoned in a grand monument on the northwest edge of the plaza, where Pustaka Bustanussalatin had installed a memorial that curated the

Figure 2.2 Example of monolith at *Thank the Nations Plaza* (Blang Padang) (source: author).

collection of boat-plaques. A large five-crested concrete wave holds a plaque that indicates the date and magnitude of the catastrophe. It also reads, in part:

> Helping hands came from volunteers, non-governmental organizations and State institutions, businesses, civil and military, domestic and international organizations. All came together to forget differences and join hands to rebuild Aceh and Nias. Thanks the World park is dedicated to the spirit of peace and friendship. [*sic*]

Additionally, a nearby monument was inaugurated on June 19, 2008, by Yayasan Harapan Nusantara Nations (Jakarta, Indonesia), a project funded by Purple Lotus Temple and Mother Samantha Foundation (both from San Francisco, US). In the shape of an obelisk, it commemorates the day of the catastrophe rather abstractly. One might imagine Banda Aceh as an urban space where monuments kept accumulating over the years without much conversation or consensus about it, the symbolic gravestones and remnants of the actions of different civil societies around the world. The practice of pinning the urban map of Aceh with different monuments was a prominent feature in the recovery process of the urban landscape of post-tsunami Aceh, providing an assemblage to the self-determination during this period as a "tsunami city." A constant in this assemblage was a memorizalization of the distinction between the time before and the time after the tsunami that marked the characteristics of this traumascape.

This is best exemplified in the inaugural exhibition in the new Aceh Tsunami Museum in late 2009. This display featured a combination of powerful images, dominated by large-format color photographs of victims and ruins, and a selection of interpretive paintings of the catastrophe. The photographic collection was organized according to three temporal experiences: *Pra*-tsunami (before), *Saat*-tsunami (during), and *Pasca*-tsunami (after the tsunami). In this exhibition, the time before the tsunami was chronologically generous. It included rare images of earthquake-damaged structures in the minutes before the tsunami wave reached Aceh, but also featured sepia photographs of Dutch colonial portraits and pictures of traditional performances and arts in Acehnese culture that are difficult to date due to their continuity as living traditions. The post-tsunami images, on the other hand, focused on the disjointed livelihoods of the city: a close-up of a broken clock with its hands frozen, marking the time "8:17"; images of overturned cars on the streets; and fishing boats wrecked on the streets. It also included architectural visions for upcoming post-tsunami features in the city, such as the Siron mass grave and the tsunami boat memorials, architecturally displaying well-thought-out designs for future spaces of memory. The disaster moment itself, *Saat*-tsunami, was expressed through a collection of images of chaotic debris and exposed corpses, often in combination. Death has been a central theme in the rebirth of Aceh, and the decomposing body remained an openly circulated image, particularly at the PLTD Apung 1 interpretation site, where a wall-to-wall display of images of corpses dominated this interpretive site (Figure 2.3, in the background). Yet, the centrality and rawness of the corpse in this exhibition stood in sharp contrast to the simplicity, sterility, and peace that could be experienced in both traditional gravesites and tsunami mass graves throughout the Aceh region. In the aftermath of the tsunami, Metro TV in Indonesia had capitalized on this "architecture of grief and aesthetics of absence" (Sturken 2007, 5), with explicit footage of the dead alongside the tally of victims, updated on the side-bar of the screen—even set to the emotive soundtrack of the Hollywood film *Pearl Harbor* on occasion.

Earlier in this same decade, the dead had held a much more invisible and controversial presence. In the early 2000s, reports began to surface about the discovery of mass graves in different towns of the Aceh province. TNI and GAM representatives blamed each other, imputing human rights abuses. Exx-on Mobil was accused of cooperating in the concealing of mass burials af-ter a particular clash with separatists. Mass graves containing the bodies of about fifty civilians were reported to exist near Bireuen, although this was

Figure 2.3 PLTD Apung I *Taman Edukasi* interpretive site (source: author).

never confirmed (Martinkus 2004, 248). As the Indonesian government issued restrictions on foreign media coverage of the separatist conflict, the accusation began to take a legendary strength. Needless to say, these graves are not mapped or included in any historical itineraries of the region, and therefore remain part of the invisible, an underground historical landscape that echoes a similarly suppressed traumascape in Bali (Byrne 2007).

As such, the time before and the time after the tsunami may not have remained distinct after all, but rather blurred or collided on certain themes. A landscape of reconstruction could also be read as a landscape of reappropriation of past narratives to be replaced with new ones. Formerly colonial-era and TNI memoryscapes were aptly replaced by tsunami-era memories, with heritage constructs playing a key role in the realization of this process. Beyond a victim of transformative forces, heritage was an instrument of construction. These efforts acknowledge that vulnerability was not attached to the material constructs themselves but to the narratives of resilience that could have emerged. The efforts of donors were channeled to navigate around this delicate balance, but not always successfully. The loudest piece of the post-disaster picture was without a doubt the Aceh Tsunami Museum (Figures 2.4

Figure 2.4 View of the Aceh Tsunami Museum from Kerkhoff Dutch cemetery (source: author).

and 2.5). A host of controversy, it was created by the Indonesian architect Ridwan Kamil as the result of an international architectural contest. When interviewed about his inspiration, Kamil stated, "For me, the tsunami museum has to reflect also the psychology that people went through during these terrible time[s]" (Bouchard 2007). A glossy poster of the competition translates the architectural vision of his abstract design concept: from above, the plan is an analogy of a spiraling sea wave, while from a distance, the building looks like a boat. The location of the museum, on the hill adjacent to the Kerkhof Peucut Dutch cemetery, also suggested this was at once a historically strategic location and a space of mass casualties.

It can be said that the museum was destined to be controversial, given that it attempted to be everything to all observers. The vision claimed to unify past and future, the religious and the secular, traditional icons and modern icons at once. A number of references to traditional Acehnese culture included, among other decorative elements, exterior paneling that recalls the intricate design of Arab *muqarna*, interior walls adorned with traditional *Saman* dancers from Aceh, and a structural layout inspired by the traditional *Rumoh Aceh* (houses on stilts). The latter was said to honor the "wisdom of traditional architecture"

Figure 2.5 Interior view of the Aceh Tsunami Museum on opening day (source: author).

that can respond to natural disasters—although there is no known connection between the evolution of the design of traditional raised houses and an awareness of tsunamis. Only now is the "soft floor" under the stilts believed to be structurally optimal in its ability to minimize tsunami damage; thus the museum's design lends traditional practices a new use-value and urgency to reinforce the place of traditional architecture. If the details of the museum would seem to be in dialogue with cultural heritage, the English names of the winning design—"Tsunami Passage" and "Atrium of Hope"—and the architectural renderings of other proposals that included Lamborghinis parked on the museum lot suggest that the architectural vision behind this project may have been significantly disconnected from the urban identity and reality of Banda Aceh. In fact, the way that the museum towers above all else in the center of the city cannot be considered a testament to traditional values. In a postdisaster city that has been likened to post–atomic bomb Hiroshima there is a stark absence of modernist and monumental structures of any scale, making the museum's building one of a kind.

The marketing of the design as a reinforcement of Achenese tradition was also challenged by the founder of Aceh Heritage Community Foundation (AHCF) in Aceh, who argued that the natural hill on which the structure was built possessed a different historical significance than its relatively recent characteristic as an "escape hill" during the tsunami. In this well-founded critique, the representative of AHCF reported how the railway station had been demolished to make room for the museum and, moreover, another building needed removal to fit the large footprint of the design (Bouchard 2007). Her concern, shared by many Acehnese, condemned this destruction of the past to accommodate a structure whose representation of Acehnese identity was questionable at best. Therefore, she openly opposed this project on the grounds that it involved the destruction of the past. BRR officials argued in response that the historic buildings to be destroyed were damaged beyond repair, even as they looked into ways of incorporating the remaining structure into the museum design. If the structures were in fact preserved underground or "in spirit," this was not reflected in the museum experience or design. The museum was also reported to have needed US$7.4 million to be completed, at a time when resources were still needed for the basic rebuilding of livelihoods in the region, so it was not clear why BRR would support its construction above all else. While many Acehnese viewed the museum as an inappropriate expenditure, the assistant manager for public facilities and buildings at BRR, Adamy Aulina, was quoted as stating that the building was necessary for symbolizing respect for the victims and facilitating the remembrance of survivors (Bouchard 2007). In a number of online and conference publications, Dr. Jonathan Zilberg, museum ethnographer and curator of the nongovernmental organization Komunitas Tikar Pandan in Banda Aceh, raised a series of moral and political issues associated with its construction, calling the new structure an immoral misuse of foreign charity and the most expensive white elephant in Indonesian history (Zilberg 2009). Zilberg argued against the construction of a new museum that undermined the existence of a legitimate Aceh museum. In particular, he was concerned with the lack of civic participation and open intellectual reflection associated with the project. Through a collection of interviews, he was able to articulate a clear resistance to the project. Komunitas Tikar Pandan Aceh, working in the sphere of arts and culture, prepared a statement that included the following critiques: "Public spaces to celebrate historical interpretation are owned by no one. It belongs to an era . . . What Aceh needs now is not a museum which is newly built, driven by the spirit of disaster capitalism and the politics of donations." Then reflecting on the price tag of the construction, Zilberg (2008) declared, "What an exuberant, arrogant, and cunning way to institutionalize Aceh's [recent tragic] past."

Joining this argument, Dr. Mukhlis A. Hamid, a lecturer in the Department of Literature and the Humanities at Syah Kuala University in Banda Aceh, presented certain conditions under which the museum would be accepted, including a clearly defined mission and educational objectives, an economically sustainable and nonprofit-based operation, and a prioritization for its construction sensitive to the adequate rehousing of all tsunami victims (Zilberg 2008).[7] Among Zilberg's diligent documentation of public voices, a woman argued in support of the construction of the museum, claiming that "if you want to learn about Aceh through its historical artifacts, documents and other aspects of its cultural heritage, you will have to go to Leiden." But Dr. Nurdin, an important local supporter for the museumification of the tsunami, championed the use of ruins as monuments to the tsunami, expressing frustration at the cleansing of the landscape. He further argued that the museum should be seen as a significant resource for tourist attraction and economic development in Aceh, a means of healing historical trauma associated with the event that is fair use of rehabilitation funds. Zilberg (2008) refers to this rationale as the "perverse logic of creating a tourist site to stimulate development through some form of global witness." Zilberg brought up an issue that was at the core of local activists' and intellectuals' rejection of the project—namely, the promotion of a culturally inappropriate form of dealing with psychological trauma and memorialization, in constrast to the role of prayer and mosques. This argument could be extended to much of the aid work and heritage construction under discussion during this period in Aceh. In Zilberg's argument, the victims themselves were the living museum and monument to this catastrophic event and, as such, they should be prioritized in the post-tsunami reconstruction process.

What these diverging views represent is the diverse ways that commemoration of the tsunami may be articulated in the context of a fear of forgetting (Baudrillard 1994, 49). There were two ways in which the landscape of memorialization and trauma was being presented in Aceh. The explicit way was articulated through iterations of disaster tourism and the commodification of the tsunami to attract an imagined audience, a heritage that finds empowerment in a narrative of destruction that is carefully articulated for specific purposes. The more covert way was through an emerging heritage assemblage that successfully escaped the centrality of endangerment, empowered instead for its role as a witness of traumatic experiences, and made sense of through the preferred spiritual and historic context of Aceh. The latter stood as an example of locally emergent heritage constructs that are left to be built and operated independently of aforementioned dominant agendas and the imposition of historical trajectories. These then become stewards to memories and narratives that explain, interpret, and preserve collective experiences of destruction, catastrophe, and decay in the context of loss, in this case, by natural disaster. In both views, what is key to highlight is the way that heritage construction implicates practices that perpetuate, resist, or adapt to the forces of destruction working against them. This is a negotiation that should be inherent to the critical study of heritage, considering the way that endangerment is rigidly embedded in established global, regional, and local heritage discourses and practices that are summoned in the work of reconstruction and the creation of heritage assemblages, the focus of the following chapter.

3 The Construction of Destruction

Despite the obvious centrality of destruction in the concerns that mediate the dominant practices of cultural heritage preservation and management, the articulation of endangerment and risk in these practices is not overtly and critically addressed. Yet, this concern is embedded in a broad range of documents, conventions, charters, and instruments, as well as the work of various regional, national, and international organizations. Mobilizations of a "heritage at risk" also signpost a heritage industry for Asia that is traditionally imagined to exist in contrast to a Western paradigm, shaping a vast and heterogeneous region in which the heritage of Aceh is perceived to be, or not to be, at risk. As such, this chapter provides a conceptual underpinning of the discussion of the ways in which heritage was constructed and destroyed on the ground in the aftermath of the 2004 tsunami in Aceh.

Risk Cartographies

In the immediate wake of the Boxing Day tsunami, the heritage of Aceh was deemed to be at risk within the overarching narrative of postdisaster urgency that was circulated globally and loudly for the broader affected region (Rico 2014). Under this assumption, a local organization supported by the Lestari Heritage Trust, the Aceh Heritage Community Foundation, addressed the role of heritage in post-tsunami Banda Aceh and Greater Aceh in exceptional detail.[8] An initial inventory of heritage was created in 2007, not simply to document the state of conservation of the monuments in Aceh but also to identify needs and allocate resources. Prior to these efforts, no systematic approach to the heritage of Aceh was adopted by the organization charged to carry out such initiatives, the state-wide conservation agency known as BP3 (*Balai Pelestarian Peninggalan Purbakala*, Archaeological Heritage Preservation Association), which has a chapter based in Banda Aceh.[9] Detailed before-and-after condition assessments were carried out by the team of architects and preservationists in AHCF, with the aim of producing a database and disseminating their efforts through a CD, a website, an inventory book, and a short documentary film. Moreover, with a team based locally, this work was updated as necessary and when permissible. The survey aimed to assess any efforts and changes associated with the conservation and preservation of sites. It also aimed to identify the parties (aid agencies, government institutions, communities) who were involved in the rehabilitation projects of the 152 sites that were identified through this initiative. In tandem, a survey of intangible heritage was carried out in February 2008, with the aim of associating heritage places with livelihood rehabilitation issues. But risk was not easy to identify in all heritage contexts. Work undertaken by the same organization with the US-based nonprofit Global Heritage Fund (GHF) in Gunung Leuser National Park had also aimed to document "heritage at risk" in the upland villages of the park. Throughout this collaboration, the Indonesian team noted a divide between the way that local villagers identified preservation needs and the expectations of the GHF, whose prospective work had preselected certain types of heritage that required assistance. It appeared that the foreign organization sought local expert validation to support a subject of preservation that had already been formalized, a "heritage at risk" that was preemptively tagged for safeguarding prior to assessment. Aceh's location on imagined risk maps made it a perfect candidate for the search for heritage subjects in peril.

Heritage-at-risk maps are now familiar mediums of communication in global heritage networks. Not atypically, CyArk, a nonprofit organization in the United States that holds a virtual, online 3-D library of cultural heritage sites around the world, communicates heritage priorities through a hazard map that depicts, in their view, "potentially devastating effects of seismic activity and climate change."[10] Another US-based nonprofit, World Monuments Fund, has instituted a "watch map" to feature the sites nominated and inscribed on the biannual Watch List. And UNESCO's Interactive Map for the World Heritage List depicts inscribed sites around the globe, with the ability to highlight sites from the World Heritage List in danger. These familiar maps show a distribution of heritage at risk across the globe that is not selected at random but predetermined by experts and their organizations. They draw their parameters from the very definition of heritage as "at risk," as an established and uniform characteristic that can be bestowed on all heritage subjects. As such, risk maps are key instruments of the rhetorical articulations of global heritage (Rico 2014). They point to hot spots of environmental instability, endemic conflict, and poor management resources as they intersect with perceived heritage significance, in a real-time scenario that aims to present and even predict a heritage that may be at risk from some kind of perceived existing or potential force of destruction. Risk maps draw their authority from selected known information from the present and the past to mobilize a future-oriented perspective and discourse. Such maps highlight the centrality of time—by way of the promotion of a sense of *urgency*—to practices and processes of heritage management.

The resulting product is a practice of mapping territories through the activity of risk assessment. Through each assessment, risk is identified and defined as a possibility based on the particular context of each heritage entity, but also informed by an expectation of the state of being at risk. As a result, it could be argued that this judgment is predominantly carried out independently of local, geographically, or temporally situated assessments of threats and vulnerabilities. Because risk maps are created using the same expert knowledge that determines heritage significance and risk factors at the global level regardless of context, the system remains hermetically sealed to any type of counter-description and therefore shows overwhelming coherence across institutions. In examining the expert networks charged with the assessment of risk in the aforementioned heritage organizations, it is not surprising to see that there is significant overlap of expertise. This is especially significant as "risk management" established itself as a unique heritage management and conservation specialty. The observed coherence therefore creates risk-averse regions that reflect current territories of heritage expertise with a gravitational center in the West. As a result, risk-prone territories are featured more prominently in less-developed nations. In part, this is expected since it could be argued that disasters and wars occur more frequently, with more intensity, and with more lasting effects in nations that are not able to manage threats adequately or consistently, or lack the resources for effective risk-mitigation strategies. However, the way that risk is defined and calculated plays an equally large part in this cartographic construction. Critical heritage debates argue that a hegemonic cultural heritage construct is able to highlight and promote only specific forms of heritage that are dominated by monumentality and whose legitimacy is defined through assessments of authenticity, integrity, and uniqueness (e.g., Labadi 2005). Within this ecosystem of norms, then, risk is defined and calculated within relatively narrow paradigms of what makes a legitimate heritage construct. As a result, the determination of heritage at risk for a traditional heritage expert may not necessarily be heritage at risk for any other stakeholder. In fact, it is tempting to propose that the condition of risk is exacerbated the further away a heritage moves from what the dominant construct promotes as heritage. This is the precarious place that the heritage of disaster inhabits, where its integrity is measured against a predisaster state but at the same time it is constructed in a new context that can give rise to new heritage value. It is a crisis of interpretation that predominantly results in an expert assessment that weighs in favor of predisaster times.

Heritage expertise relies on disciplinary doctrines to support risk assessments. For example, modern conservation approaches state that a failure to distinguish old construction from new reconstruction increases the vulnerability of heritage subjects. Living heritage is also placed at odds with risk mitigation—for example, practices of pilgrimage are considered to be risk-inducing as they lack consideration of the carrying capacity of heritage risk-prone landscapes (Rohit Jigyasu, personal communication). At the same time, heritage experts themselves have openly disagreed on the state of conservation and fate of different heritage sites, although this is rarely formally articulated in these terms (Meskell 2014). Yet risk takes a prominent role in the construction of a normative framework for the field of cultural heritage preservation, universally applying the inherent perception of endangerment or risk across all material culture regardless of its particularities (Rico 2014, 2015). Risk assessment methodologies thus attempt to rid the problem of its subjective nature. Focused prominently on materiality and addressing threats and hazards as decontextualized factors, these methods often disregard contextual cause and effect. They may erase, for example, the difference between natural and man-made destructive factors (Benarie and Druzik 1992) or rely on abstract models of deterioration and destruction in order to establish long-term statistical losses (Druzik 1992), although more recent initiatives aim to address risk assessment more holistically. Methodologies reinforce the construction of an abstract and transferable risk assessment through the creation of authoritative lists that formalize an antithetical relationship between heritage constructs and processes of destruction (Harrison 2016; Rico 2014, 2015).

Heritage is constructed against specific perceptions of processes of destruction and their effect. Heritage definitions may change to keep up with the evolving understandings of heritage construction at different points in history and in different contexts. But despite accommodating widely discrepant meanings (Larkham 1995, 85; Lowenthal 1998, 94; Samuel 1994, 205), the centrality of destruction remains nonetheless intact. Whereas discussions of what constitutes heritage have been a key focus of enquiry throughout the late twentieth century and early twenty-first century, the emergence of the concept rarely highlights the significance of actual and perceived destruction for the articulation and formalization of heritage subjects, methods, and practices. In fact, a chronological examination of definitions of heritage through documents issued and circulated in the history of heritage-related disciplines reveals overwhelming homogeneity regarding the question of change. Preservation by definition is the management of change (Pye 2001), and in this mission all change is inconvenient and no uncontrolled change is permissible in principle. The state of ruination has been a distinctive anxiety reflected in the dominant heritage doctrine. It is not surprising that the experience of ruins as heritage developed in the West in the 1800s (Burke 1969), a shift away from the anachronistic attitude of the Middle Ages, when ruins were assimilated as "mirabilia" (marvels) without questioning origins or difference. An admiration of the ruin reemerged in the literature of the late nineteenth and twentieth centuries (Chamberlin 1979; Harvey 2001), and it could be said that still remains today, although the distinction between romantic ruin and derelict structure remains mystified (DeSilvey and Edensor 2013). What is clear is that ruination in this context comes into conflict with the concept of integrity, defined by UNESCO as: "the physical fabric of the property and/or its significant features should be in good condition, and the impact of deterioration processes controlled" (UNESCO 1972b, 23), while relationships and dynamics essential to their distinctive character should also be maintained, a requirement that has demanded significant explanation (Stovel 1995, 2007) and sets a high bar for the relatively limited tools of cultural heritage preservation.

Heritage and Destruction

When is heritage at odds with processes of destruction, then? The establishment of heritage at risk is determined by an examination of factors of vulnerability and threat. Vulnerability in cultural heritage is associated with debates around the existence of inherent heritage value, a heritage "aura" in material culture: a characteristic that resides in genuine artifacts and cannot be reproduced. But the existence of an "aura" has been challenged as a product of discursive practices (Shanks 1998) and perceptions of the past (Holtorf 2001, 2013; Holtorf and Schadla-Hall 1999). Moreover, the recognition and advent of value-based conservation and management of cultural heritage (Australia ICOMOS 1999; Mason and Avrami 2002) opened the possibility that, given the coexistence of multiple values, heritage may be at once *at risk* and *not at risk*—depending on which values are considered to be at risk and what factors affect them in particular. The second fact that needs to be determined in the establishment of risk to heritage is what factors affect heritage values—specifically, what hazards actually or potentially affect the integrity of heritage places negatively. These factors have been well known in heritage practices at local, regional, and international levels as they are keystones in the dominant heritage discourse, but are too frequently used to generalize the challenges of heritage preservation and management rather than providing a nuanced value-based assessment of context. Factors include phenomena from localized conditions, such as temperature and relative humidity, to natural disasters, destruction by war, and intentional destruction for any purpose. There has also been growing acknowledgment that issues internal to conservation management itself may constitute a threat to the integrity of heritage under certain conditions.

This harks back to early conservation debates that pitch the patina of age against the original state of creation as the creator intended. In this formative debate for the discipline, Eugene-Emmanuel Viollet-le-Duc (1858–68) defended the *original* state and sanctioned later additions to buildings, adhering to early restoration traditions that sought to maintain or improve a building's image in a way that follows the wishes of the owner (Sitwell and Staniforth 1998; Sixth International Congress of Architects 1904). Its counterphilosophy, developed by John Ruskin, embraced destruction as part of the life history of buildings and objects, approving the belief in the equality of all layers of history and disapproving restoration practices (Ruskin 1849). In Ruskin's view, restoration was ultimately attainable and equivalent to destruction, describing the dominant nineteenth-century practice as one that creates a feeble and lifeless forgery (Society for the Protection of Ancient Buildings 1877). Despite the general dislike of age and decay that Lowenthal (1985) has described extensively, a distinction between "living" and "dead" heritage has been recognized and expressed since the early twentieth century (Sixth International Congress of Architects 1904), although a full appreciation of living heritage that challenged the assessment of destruction significantly would not be pursued until the end of the twentieth century. The definition of what constitutes destruction of heritage was integral to the direction that the discipline of heritage management would take and remains an active topic of debate among heritage experts, practitioners, and scholars.

Therefore, as threats and vulnerability come together to determine the status of heritage as being "at risk," a third and powerful weakness emerges in this assessment: the determination of the way in which threats affect heritage values can only be made subjectively, relative to the observer's perception of change and the observer's ability to rank the degree of change against an agenda that sets standards for the acceptability of change. In principle, and in actuality, heritage risk assessment should take as a starting point the assertion that all heritage is at risk, and then justify which heritage is more at risk than others to define preservation and management priorities. This would, of course, require revealing not just the instruments used in decision making but also the agendas that give power to a perception of universal heritage value and global heritage objectives. Historically, these objectives have origins in the Western hemisphere in the experiences of the destruction of postwar landscapes in the first half of the twentieth century. Postwar destruction of traditional urban environments through redevelopment of cities and the expansion of urban, suburban, and freeway developments caused a reaction against excessive transformation of the human landscape and the consequent loss of reminders of the past, a reaction that was translated into an overwhelming support for the emergence of the conservation movement. Interest in cultural and intellectual reconstruction efforts gave rise to the United Nations Educational Scientific and Cultural Organization at the end of World War II (Valderrama 1995), and subsequently to a powerful framework devised to safeguard it. As part of this framework, heritage concerns became internationalized and proliferated through specialized cultural heritage organizations that responded to a growing need to tackle the complexity and scale of emerging heritage issues: the International Council of Museums (ICOM, founded 1946), the International Centre for the Study of the Preservation and the Restoration of Cultural Property (ICCROM, founded 1956), and the International Council on Monuments and Sites (ICOMOS, founded 1964), to mention the earliest and most influential initiatives. The landmark for UNESCO's work against the forces of destructive change was the successful rescue of the monuments of Nubia, threatened with submersion in the construction of the Aswan High Dam. The request by the United Arab Republic (modern Egypt and Syria) mobilized international heritage cooperation for a thirty-year project of cultural heritage preservation (UNESCO 1960). The completion in 1980 of the first UNESCO campaign, "Victory in Nubia," consolidated a coherent program for heritage preservation that had been envisioned at the founding of the organization, and joined other high-profile projects: the work on flood-damaged Florence and Venice (1970), followed by the restoration projects in Borobudur (1973), Mohenjo Daro (1974), and Bamiyan (1974).

The definition and approaches to preservation and management of cultural heritage take shape through institutional, international, and national documents that reveal a growing concern with standardizing the principles of preservation and management of heritage hinging on and transmitting a standardization of specific attitudes to change. The Convention for the Protection of Cultural Property in the Event of Armed Conflict (UNESCO 1954), whose preamble establishes an assertive recognition that "cultural property has suffered grave damage during recent armed conflicts and that, by reason of the developments in the technique of warfare, it is in increasing danger of destruction," resonated with the aims and postwar origins of UNESCO and the dominant heritage framework established therein.

UNESCO thereafter continued to produce recommendations that defined the application of these principles in the context of archaeological excavations (UNESCO 1956), landscapes (UNESCO 1962), illicit transnational trade (UNESCO 1964), and public and private works (UNESCO 1968), to mention a few. A landmark work in this production of texts is the 1972 Convention Concerning the Protection of the World Cultural and Natural Heritage (UNESCO 1972a), whose preamble asserts the imminent destruction of heritage:

> *Noting* that the cultural heritage and the natural heritage are increasingly threatened with destruction not only by the traditional causes of decay, but also by changing social and economic conditions which aggravate the situation with even more formidable phenomena of damage or destruction. (UNESCO 1972a, Preamble)

Among the structural changes that are asked of signatory state parties, the convention demands the development of scientific studies and technical studies and research toward counteracting the dangers that threaten its cultural and natural heritage (UNESCO 1972a, Art 5). On this foundation, the global heritage model put forward by UNESCO established the identification and valorization of cultural heritage by consensus through the World Heritage List model, starting in 1978 and growing ever since as more cultural, natural, and mixed heritage sites are attached to the universal vision that was articulated in the 1972 convention.

In 1964, the adoption of the Venice Charter called on nations to agree to the principles guiding the preservation and restoration of ancient buildings, "with each country being responsible for applying the plan within the framework of its own culture and traditions" (ICOMOS 1964, Preamble). However, in the relatively short period of existence of the World Heritage List, an unequal distribution and overrepresentation of certain geographical areas and certain types of heritage emerged, pointing to a lack of harmony between international and local constructions of cultural heritage and its management. The Global Strategy (Labadi 2005; UNESCO 2007) and the Gap Report (ICOMOS 2004) acknowledged and described this imbalance, but the results of these efforts did not affect the model structurally (Rico 2008). However, scrutiny of the outcomes of the 1972 convention revisited and strengthened earlier critiques of its Eurocentric roots (Cleere 2000). It also highlighted the marginalization of types of heritage that do not conform to the dominant heritage framework that has been discussed at length as the product of an authorized heritage discourse (L. Smith 2006). A history of contested heritage and its uneasy relationship to the World Heritage List model put into question the relevance of the promotion of universal heritage values (Blake 2000)—in particular, how they apply to perceptions of destruction.

Overwhelming evidence of the intentional destruction of cultural heritage under the instruments of the dominant heritage discourse itself has raised awareness of the effects of such an inflexible framework on the diversity of cultural heritage constructs worldwide. Perceived conflicts between coexisting historical, scientific, and spiritual values and interpretations have been used as rationale for the destruction of heritage in the name of authenticity (i.e., Ayodhya/Babri Masjid; Shaw 2000), rights (i.e., Bamiyan Buddhas; Meskell 2002), and conflict (i.e., Stari Most bridge; Silverman and Ruggles

2007). These globally debated cases suggest that landscapes of destruction are mobilized and authenticated selectively, in accordance with covert agendas of inclusion and exclusion from legitimate heritage assemblages. A clear example of the exclusion of heritage associated with destruction is the mobilization of a "negative heritage" (Meskell 2002) and the undeniable scarcity of globally recognized cultural heritage sites that openly represent dissonant interpretations and express openly the suppression of historical, spiritual, or other values. Because of this, the dominant global model has been criticized for engaging in an active discouragement and careful manipulation of conflict (Meskell 2002; Turtinen 2000), arguing that those constituents that are included follow a careful process of selection that strips away dissonance (Chadha 2006; Rico 2008; Turnbridge and Ashworth 1996). Such decontextualization from the forces behind the creation of heritage places and heritage significance has been discussed as a process of reterritorialization (Breglia 2006) that conforms to an established heritage "grammar" (Lafrenz-Samuels and Rico 2015). Global mapping is an essential instrument that *teaches* member states to have world heritage (Turtinen 2000, 3), including in particular the mapping of "heritage at risk."

Risk Value

Value-based approaches that take into consideration the human contexts in which cultural heritage manifestations exist and change have been central to the management of heritage since they were proposed by the Australia ICOMOS Charter for the Conservation of Places of Cultural Significance (the Burra Charter; Australia ICOMOS 1999). First adopted in 1979, the Burra Charter has had long-lasting impact in heritage management at the global scale as it defines cultural significance to consist of "aesthetic, historic, scientific or social value for past, present or future generations" (Australia ICOMOS 1999, 2). It thus provides a systematic approach to conservation planning that separates the assessment of significance from management decisions. This reconceptualization not only reflected an acceptance of the changing and multiple interpretations in the meaning of heritage but also encouraged the inclusion of marginalized perceptions and constructions of heritage in management and practice (Byrne, Brayshaw, and Ireland 2003). As a learned, rather than innate, abstraction, it has been argued that values depend on the particular cultural, intellectual, historical, and psychological frames of reference held by the particular individuals or groups involved in determining them (Lipe 1984). However, less effort has been put into the design of adequate methodologies to capture the array of perceptions and interpretations that value-based assessments require. Moreover, Article 80 of the World Heritage Convention's Operational Guidelines states that values are authoritative according to "the degree to which information sources about this value may be understood as credible or truthful" (UNESCO 1972b, 21), conveniently shifting legitimacy to some form of recognized expertise.

The deployment of an unarticulated "risk value" is an example of this problematic practice, as a value that cannot be measured but resides instead in a situated perception that gives heritage and its preservation a sense of urgency and importance. Risk value, however formative for cultural heritage, is not an articulated value in statements of significance that are critical to heritage discourse, but is utilized prolifically in heritage rhetoric to justify the objectives and means of the discipline (Rico 2015). This is surprising when considering the academic and institutional efforts that have been put into defining the concepts of authenticity, integrity, and significance, foundational concepts that hinge on perceptions of change. At the turn of the twentieth century, Alois Riegl (1996) problematized change and destruction through discussions of a cult of age value that condemned violent destruction of monuments through the hand of man as an interference with nature's lawful activity of disintegration. In principle, this attitude permitted calm and lawful continuity of the disintegrating effects of nature's forces, but opposed sudden, violent destruction. In this view, the cult of age value directly disapproved of the preservation of monuments, as this would interfere with the identification of age, revealed in imperfection and a lack of completeness. Destruction enjoys a conflicted position in the realization of age value. Growth into decay is seen as part of an expected and welcome natural activity that should not be obstructed as long as it occurs gradually, but on the same note, a rapid destruction is seen as unnatural. Though the destruction by a natural phenomenon is not addressed in this dichotomy, this perspective puts into question the normative and privileged place that "heritage at risk" takes in heritage discourse.

Ontological questions surrounding the essence of heritage have emerged more recently in relation to its endangerment. Rodney Harrison (2013) argued that a sense of uncertainty, vulnerability, or risk informs the relationship of heritage and modernity as a philosophical and political concept. Heritage stands in an uneasy relationship with time, haunted by the idea of decay, while recognizing that progress and obsolescence are inevitable, and producing what he calls "opposing sentiments" (2013, 26) that involve a desire to be liberated from—but also conserve—the past. It is an ambiguous relationship characteristic of modernity that reveals itself more obviously in the modern concept of risk. A closer examination of this concept emphasizes that risk is a product of perceptions of vulnerability, a key instrument of globalization that supports a modern industrial society (Beck 1992, 2002; Douglas 1992, 15). Mary Douglas's (1966) problematization of the risk construct and its management is associated with a transgression of boundaries and the idea of being "out of place." Harrison (2013) suggests that this transgression is linked to anomalies within classificatory systems, treated with distrust as potential sources of social disorder and threat—the globalization of doubt (Beck 1992, 21). The listing and ranking of heritage through risk further entangle practices of heritage recognition with a globalizing endangerment sensibility (Harrison 2016).

The attribution of risk value to heritage, then, constitutes an attempt to colonize the future in consideration of the consequences of human action (Beck 2002). But Beck convincingly argued that this should be seen as a manufactured uncertainty that relies on the articulation of global shared threats that are questionable and difficult to quantify (Beck 2002, 40; A. Lakoff 2007, 7; Vidal and Dias 2016). In this challenging task, technical expertise is charged with the calculation and management of the risks that it generates: an uncertainty about the articulation of uncertainty as it is noted in "heritage at risk" models nonetheless results in a prolific dissemination of the "heritage at risk" rhetoric for the promotion and justification of technologies of expertise that claim to understand, identify, and manage risk in heritage constructs (Rico 2014, 2015). Outside of heritage debates, a closer examination of the articulation of risks reveals that it requires a common ground of shared values in order to rank dangers—a requirement of risk assessment—based on agreed-upon criteria (Douglas and Wildavsky 1982). Arguing that "common values lead to common fears," Douglas and Wildavsky (1982, 8) also add that the level of acceptability of a given level of risk is a political question.

More focused research has been recently dedicated to understanding the practices and policies at work associated with the fabrication of a "heritage at risk" model within the operationalization of world heritage (Meskell 2013). This ethnographic work reveals the political transactions that are involved in challenging or supporting heritage expertise in their determination of priorities and urgency. The end result is the inscription of sites into the World Heritage List and, in particular, the World Heritage in Danger List. In a dedicated study of the mobilization of "states of conservation," Lynn Meskell (2014) further demonstrates that the determination of which sites are recognized to be under threat and in need of emergency action does not necessarily follow the assessment of experts but rather is used to support state ambitions that can be conveniently propped by heritage concerns. Meskell has called this politicking a "destruction happening daily," while heritage concerns focus instead on conflict and war zones.

As I discuss the example of Acehnese postdisaster heritage debates that run through this book and in particular in Chapter 4, risk is calculated and defined by a range of experts who are temporally, spatially, and ideologically situated, privileging abstract expert systems over local forms of knowledge (Giddens 1991, 29–32, quoted in Harrison 2013, 27). Experts may then rely on seemingly disparate types of events that fit into the framework (A. Lakoff 2007), as dominant heritage rhetoric encourages reliance on a discourse that, at it's most extreme, equates the loss of heritage with the loss of a human future. In postdisaster heritage landscapes, and in particular in disaster-prone territories, this rhetoric thrives, while standing in opposition to the locally framed risk of disappearance of emerging heritage forms.

Asia "at Risk"

There is no compelling rationale for anticipating risk value within certain territories, other than granting that this anticipation is rooted in a dominant interpretation of destruction from a predominantly Western perspective. According to this expectation, heritage in Asia in the aftermath of a disaster must be, by default, "at risk." But an assertive deconstruction of heritage discourse in the past few decades has demonstrated that the heritage construct may exist without a need to preserve it—that is, heritage may flourish, indeed, heritage value might even be understood to *increase*, in harmony with its destruction. Asian heritage constructs hold a pivotal role in this discursive shift.

Preservation efforts across the vast territory that is Asia have been dominated by two concerns. On one hand, there is a noted preoccupation with the loss of the historic fabric in the process of rapid urbanization and modernization of urban centers (i.e., R. Jones and Shaw 2006). A United Nations Economic and Social Commission for Asia and the Pacific (ESCAP) report noted in 1993 that "the speed and scale of urbanization in Asia now poses major challenges to the prospects for sustainable urbanization in the region" (Howe and Logan 2002, 2). On the other hand, the active commodification of traditional practices for the purposes of tourism (Hitchcock and King 2003; Winter and Daly 2012) seeks to create an authentic Asian experience that can result in a stagnation of performances that affects the self-determination of an indigenous authenticity. In Indonesia, the changing attitudes to colonial heritage between the rule of Sukarno and that of Suharto resulted in inconsistent preservation practices (Coté 2002). It is now recognized that under Suharto's New Order, the reconstruction of the past became more purposeful, effective, and militarized (Reid 2005c). In addition, a state-sponsored "touristification" in the New Order used cultural tourism to enhance its political legitimacy and boost economic growth, a phenomenon that has been particularly observed during the *reformasi* in Bali (Picard 2003), when broader ideological interests were revealed (cf. Winter 2007, 16).

Heritage charters in Asia reflect these concerns, stimulated by an interest in accommodating greater cultural diversity in the way heritage is constructed, understood, and managed. They emerge in response to the growing number of case studies that suggest that Western-influenced international heritage charters are not always transported adequately to Asian regions (Howe and

Logan 2002, 6). Early expressions of a regional approach were mapped and disseminated in the Bangkok Charter of 1985, which proclaimed resistance to the restrictions imposed by the Venice Charter on the restoration of Buddhist monuments in Sukhothai (Byrne 2004). A uniquely Asian approach to heritage preservation, however, was not popularized until the 1994 Nara Document on Authenticity (ICOMOS Symposia 1994), a key document in this trajectory as it reflects a shift in value assessments that acknowledge the source of authenticity located in the living community most closely identified with its origins. The twenty-first century has seen a proliferation of charters and principles that cater to Asian heritage and aim to draw a distinction between Western heritage management and the principles that govern the identification and protection of non-Western heritage.

For example, the 2000 ASEAN Declaration on Cultural Heritage in Bangkok recognized a need to identify, recognize, and protect cultural landscapes. The Hoi An Protocols, drafted in the same year and revised in 2005 (Engelhardt and Rogers 2009), promote professional guidelines for assuring and preserving the authenticity of heritage sites in the context of the cultures of Asia. Drawing considerably on the Burra Charter and the Nara document, the China Principles recognize the impact that tourism has on the presentation and restoration of heritage in Asia (Qian 2007). The robustness of an "Asian" model is informed by the role of local communities in the construction and transmission of heritage, recognizing and integrating tourism development agendas in heritage management, and appealing for methodological improvements that allow the incorporation of these perspectives in the regional work of heritage. It is an approach that sets forth a cultural heritage that is dynamic and continuous. However, these documents have also been accused of reinforcing Eurocentric understandings of culture and landscape through their discourse (Qian 2007, 258; Winter 2007, 14). For example, the recognition and inclusion of intangible heritage within global heritage instruments presented an internal contradiction: it stems from and forms part of a static character of preservation, while representing the dynamic processes of landscape development (Haber 1995). An Asian legacy in heritage studies informed and empowered the articulation of approaches to non-Western heritage that had a strong impact at the global level, wherever resistance to Western heritage hegemony sought to shift the directional focus of global discourse.

Articulated in particular in heritage debates since the 1980s, dedicated studies into an "Asian perspective" played a role in the critical examination of the role of cultural heritage management in the deployment of Western imperialism in non-Western territories (Byrne 1991). Irreconcilable differences between Eastern and Western perceptions of heritage conservation were explored, as the West became synonymous with the objective reality of monuments while the East was said to focus on the *genius loci*, the spirit of a place (Wei and Aass 1989). Moreover, the relationship between landscapes and memory was distinguished in Western and non-Western contexts, the latter justifying the significance of cultural landscapes as an analytical category (Ferguson and Anyon 2001; Küchler 1993, 85–86). The articulation of these dichotomies challenged at once the foundational principles of authenticity and integrity as well as the channels of authority that award dominant heritage constructs their operational power. Among widely debated examples was the preservation of the Ise Shrine in Japan, famous for its reconstruction tradition that involves cyclical and complete rebuilding of the structure (C. Adams 1998). Denis Byrne's (1995) early study of the preservation of Thai Buddhist *stupas* and, more broadly, heritage management in Thailand, brought to light the value assigned to new fabric in accordance with Buddha's final lesson on impermanence, which states that decay is inevitable. By contrasting preservation principles and practices in a context where abandonment, decay, and impoverishment are continuously balanced against the process of maintenance and restoration, his work questioned a privileging of the original form in the context of Thai local religious practices (Byrne 1995, 274–275). Of particular relevance to the study of restorations of religious heritage, Byrne's work empowers a critique of the dominant heritage framework that proposes that the past may be seen as a renewable or "at least a reproducible" resource (Holtorf and Schadla-Hall 1999, 234). Anna Karlström (2005) continued this debate on the relationship between modernist ideas of preservation and the context of Buddhism through her work in Laos, stressing that preservation is an active process of materialization, where materiality is anything but an empirical reality (Buchli 2002, 14–15). Her research explores the notion of impermanence and the value of decay especially as they pertain to a spiritual and cultural understanding of time, challenging established ideas of integrity through an argument that suggests the production of heritage can be situated in any temporal framework. Perceptions of time in Buddhism challenge chronological constructs that are instrumental to heritage, such as the idea of something being "archaeological," "the past," and, particularly, "decay." This has significant implications for the conceptualization of destruction.

An exploration of heritage sites as "living" places added to the de-privileging of the centrality and authority of a strict definition of heritage as material culture embedded with heritage value. Miura's (2005) work at Angkor Wat in Cambodia, for example, focused on the significance of studying this site, which exists as heritage and sacred site at the same time, and thus points to the inadequacy of modern preservation models, which focus predominantly on tangible factors. "Living heritage" is discussed under a different genre of vulnerabilities—for example, the de-sacralization of sites through tourism and commodification (Miura 2007). It was not until 2001 that the UNESCO Masterpieces of the Oral and Intangible Heritage recognized the existence of an intangible heritage as "people's learned processes along with knowledge, skills and creativity that inform and are developed by them" (UNESCO 2001, Definitions). The 2003 Convention for the Safeguarding of Intangible Cultural Heritage (UNESCO 2003) recognized the need for more strict regulation in the ways in which UNESCO itself deploys principles and methods of heritage preservation, promoting indigenous curatorial and preservation approaches as subjects under protection. Through these efforts, the gap between Western and Eastern conservation philosophies identified two decades before could be said to have been closed, but not without resistance (i.e., Bortolotto 2007), ocasionally reflecting the unwavering grip of Western expertise on heritage discourse (i.e., Baillie and Chippindale 2007). Although Asian heritage experiences cannot be said to be the single factor for a growing consideration of other marginalized non-Western heritages worldwide, it certainly constituted a significant part in the eye-opening resistance against dominant discourses and practices. But at the same time, a view of non-Western heritage as a monolithic model may have been unintentionally promoted.

The creation of a working category for non-Western heritage under the banner of Asia is problematic. An association of Asian heritage with intangible values and "living" practices invariably marginalizes a heritage that is solely defined through its distinction from, or rather opposition against, Western principles of cultural heritage management. Many of the experiences of heritage in Asia remain to be revealed and articulated in this approach (i.e., the "popular supernatural" that continues to be marginalized; Byrne 2004, 19). A dichotomous framework that has forced a conjured "Asian" construct forces Asian heritage to perform as a homogeneous category. This can be as effective at obscuring local perceptions and experiences of cultural heritage as the dominant globalizing discourse was at ignoring them. Parallels can be drawn with the "Asian controversy" (Erni 2008), which questions the way that indigeneity can be identified in the region, considering the widely discrepant ways in which the history and effects of European colonialism took place in different contexts (B. Kingsbury 1998). The notion that a unique Asian heritage approach can be imagined and identified is a practice of categorization that is restrictive. Instead, notions of Asian-ness as a coherent discourse should be seen as potentially damaging to diversity as notions of universality proposed through the dominant Western heritage discourse.

Indonesian Heritage

Frits Van Voorden (1999) noted that in Indonesia the traditional selection and classification of monuments actually have a detrimental effect on their potential conservation, proposing instead to look for "Asian ways" of viewing space. The dominant approach, as put forward by experts in ICOMOS

Indonesia (Affandy and Soemardi 1999), acknowledges the emergence of a heritage preservation ethos that is embedded in the colonial histories of the nation. Dramatic changes in the urban life and landscape of Indonesia at the beginning of the twentieth century had transformative effects on the rights to land ownership in Indonesia through the Agrarian Act of 1870 (Rossabi 1999), accompanied by the expansion of the railway network in Java and Sumatra, the failure of the Dutch Ethical Policy of native welfare (Gin 2004, 489), and the decentralization of municipalities in 1905 (Hartono 1999, 79). In the nineteenth century, colonial rulers had exhibited little interest in the monuments of the civilizations they had subjugated (Anderson 1993, 183); however, the early part of the twentieth century saw a growing interest from the Dutch Indies government in preserving the archaeological sites in the country, an exercise that has been identified as a colonial appropriation of monuments with the aim of legitimizing colonial rule (Anderson 1993; Glover 2003). The idea of an "Indonesian Style" in architecture emerged at this time as a reaction to the neoclassical architecture of European monumentality used to demonstrate the extent of Dutch colonial power since the nineteenth century (Hartono 1999; de Vries and Segaar-Höweler 2009), although Hartono points out that the individuals behind this movement were Dutch-educated, calling into question how indigenous an "Indonesian" style really was. By 1913 the Dutch government established an institute of antiquities, *Oudheidkundige Dienst* or O.D. Through inventories and recommendations, this institute catalogued sites throughout the country, while its director, Dr. Bosch,[11] promoted the idea of a *genius loci* of the Indonesian people (Ambary 1999, 11).

In 1951, after Indonesian suzerainty was assured, the responsibility for archaeology fell to the Archaeology Service of the Department of Culture,[12] charged with research and restoration work programs and, with the help of UNESCO, the restoration of Borobudur. In 1992, the Indonesian government passed the Cultural Protections Act, UUBCB No. 5: 1995, which extended broad protections for all monuments, sites, and cultural artifacts against theft, destruction, or any activities that might threaten the cultural heritage property of the country. The Department of History and Antiquities has since developed restoration projects at a regional level and has also assisted in funding overseas projects, such as the restoration of some of the temples at Angkor Wat in Cambodia. The current legal basis for conservation in Indonesia is UU No. 5/1992 on Cultural Heritage Assets (*Benda Cagar Budaya*), which evolved from the Dutch colonial Monumenten Ordonantie Stbd No. 283/1931. Addressing the absence of grassroots heritage voices, it has been suggested that a fear of being identified as counterrevolutionary in Indonesia tampers with the popular voice even in heritage matters (Rachman 1999). In addition, conservation concerns are politically sensitive, particularly when dealing with the remains of the Dutch colonial era, or remains of nonindigenous ethnic groups, such as the Chinese. Other critics have described conservation efforts in Indonesia as a luxury that pays homage to past legacies (Soemardi 1999). The heritage conservation movement in Indonesia was therefore initiated and formalized by nongovernment organizations and private interests in big urban centers like Jakarta, Bandung, Semarang, and Surabaya, and conservation practices remain a relatively new phenomenon (Affandy 1999, 172). Indonesia's "New Conservation" movement (Cowherd 1999) identifies conflicts between established global understandings of heritage and the particular context of Indonesian culture. Indonesia presents a context for the practice of conservation that is in principle significantly different from a Western context. On the one hand, the monumental past, as it is represented through sites such as Borobudur, Prambanan, Sriwijaya, and Trowulan, reflects a distant past whose cultures are all but gone. On the other hand, Indonesia reflects a collection of hundreds of distinct living cultures, many of which are still closely associated in their various cultural performances and traditions that reach back hundreds or thousands of years. This distinction suggests that while the principles of the Venice Charter may be adequately applied to the conservation of Indonesia's monumental past, it offers no guidance for preserving the majority of Indonesia's cultural heritage given the limitations outlined earlier—namely, the need for community agency (Cowherd 1999, 3). Robert Cowherd highlights a symbiotic relationship between material culture and living culture, where the local communities who use their built heritage are arguably more valuable as a cultural asset than the buildings themselves. Thus, an uncritical application of Western standards "might lead to the separation of a valuable physical culture from its living culture" (ibid.).

In particular, emerging ethnographic approaches have considered that cultures have long had their own methods of preserving their cultural heritage beyond the development of scientific methods and principles that have been used to justify different forms of retention and intervention by the West (Kreps 2006, 458). Cristina Kreps has convincingly argued that non-Western models of museum and curatorial practices should be observed as unique cultural expressions. Kreps (2005, 4) has demonstrated that a local term for heritage in Indonesia, *pusaka* (although some informants used *pustaka*), reflects an understanding of cultural heritage that is both tangible and intangible. Her ethnographically informed study of museums in Kalimantan demonstrates why it is difficult to translate locally constructed uses of heritage into an overarching heritage framework (Kreps 2005). She argues that *pusaka* generally translates into "heirloom" but in fact represents three related meanings: it stands for the English analogy for heirloom, for personal inheritance, and for inheritance that is of special value to a community, one that cannot be disposed of without specific common descent—the latter being analogous to heritage in the sense that it describes something that comes to be possessed as a result of one's natural situation or birth (Soebadio 1992, 15). *Pusaka* plays a significant role in the transmission of cultural knowledge and traditions over the course of generations. The point that Kreps makes through her work is that virtually anything can become *pusaka*, but not everything that is inherited becomes *pusaka* nor can objects be created to be *pusaka*—a thing must become *pusaka* in the course of its social life (Kartiwa 1992). A key theme in Kreps's (2003) thesis is that, like *pusaka*, all heritage constructs need to be integrated into larger social structures and ongoing social practices in order to preserve the socially and culturally constructed values and meanings that are contingent on specific contexts. The social and cultural dimension of Indonesians' relationships to material culture has thus been highlighted as a priority in curatorial heritage work (Damais 1992; Moses 1992). As an example, Soekmono (1992, 52) argues that the concept of *pusaka* refers to the ability of objects to possess an intimate, spiritual connection to their owner or custodian, in such a way that if *pusaka* "were in danger of theft, damage or destruction, he [the owner] would do everything in his power to save it." Traditional practices of preservation and curation of this construct are therefore adaptable and can undergo changes in response to other social and cultural changes (Kreps 2006, 468). A need to acknowledge the social and cultural dimensions of people's relationships to objects has been echoed by many who champion the need for a new approach to heritage authority.

Indonesian Law 11/2010 on Cultural Property in 2010 was meant, or expected, to acknowledge and address the lack of consideration for the value of community involvement that has been identified as key to all modern approaches to heritage management, but in particular those approaches that condemn the separation of community and heritage object, such as the one discussed earlier. However, leaning on the authority of regencies without any capability to implement such a high order, this new law is more in line with the top-down approach that underlines all heritage management and stewardship in Indonesia, including the involvement of UNESCO, which has historically been tied to debates about "classical"/high culture being preserved over vernacular/low culture (Salazar 2014; Wall and Black 2004). In this dichotomous context, the official interest in Aceh remained minimal, left to the motivated interest of individuals and their grassroots organizations. These local experts jump-started a unique debate through their privileged position as disciplinary experts and local practitioners. Despite a global mobilization of a rhetoric of endangerment that would find a fitting target in postdisaster heritage in Aceh, the heritage debates that emerged in this period instead reflected a much more nuanced and strategic appropriation of presumed "Western" and "non-Western" attitudes to heritage construction and destruction.

4 An Ethnography of "Heritage at Risk"

In Aceh, discussions of *pusaka,* a key concept considered in the previous chapter, also informs the construction of "heritage at risk." Discussing the perception of an alternative approach to cultural heritage, a leading local heritage expert contrasted the perception of heritage in Aceh as distinct from that in Europe, where "they are obsessed with objects." But, he continued, in Aceh it is very difficult to focus on objects. The destruction associated with the tsunami gave adequate justification to this claim. For the Acehnese, heritage is *pusaka.* "We don't discuss authenticity in the same way," he explained, and "neither do we follow the 'Asian model' for heritage, where materiality follows a cycle of renewal." Particularly after the tsunami, the Acehnese were very concerned with keeping their cultural values intact, afraid that *kampung* (village) values would be lost by the tsunami's destruction, as they had been in the past, when young people lost their traditions and therefore their ability to rebuild society. "*That* is what is at risk," I was told (Dr. K., personal communication, 2010). This would not be the only time that the identification of risk would require a more nuanced understanding of perceived destruction and agency. But this brief conversation anticipated a substantiated skepticism about the definition and identification of what is damaged and "at risk" in the heritage of Aceh. It suggested that a more holistic analysis was needed in order to capture the blending of dominant preservation approaches and independently crafted appropriations of heritage discourses in creative and contextual ways. This chapter aims to offer such a perspective, examining with an ethnographic sensibility the construction of a post-tsunami heritage as a process that intersects three key factors: the influence of the formalization of a "culture of disaster," the Islamization of tangible cultural heritage places and memorials, and the significance of a chronological sensitivity in the construction of heritage subjects.

Heritage Ethnography

An ethnographic sensibility in heritage research moves the focus of enquiry into heritage away from the degree of destruction that postdisaster Aceh may feature, and away from the concern with extinction, toward the way that place and people's lives are intertwined so that narrative, experience, and the urban and natural environment are inseparable. Ethnographic heritage research, by its very nature and objective, aims to extract the same type of complexity. Ethnography not only enriches heritage studies as a methodological avenue for the documentation of holistic and situated heritage values (Filippucci 2009, 320; Hollowell and Nicholas 2009) but also remains underestimated as an archival resource of the formation and operation of heritage values in the past. Ethnographically documented narratives of people's experienced landscapes offer details of landscapes embedded with memories and narratives that are able to be transmitted, recreated, and "revived" in the same way that heritage is claimed to operate. For example, in *Ilongot Headhunting* Renato Rosaldo (1980) demonstrates that ethnography stands to gain considerable analytical power through close attention to historical processes, documenting the construction of essential time markers that define and distinguish the past from the present through collective memories. Moreover, his subjects' accounts of the Japanese invasion of the Philippines in 1941–1945 highlight the significance of known space in the experience and recollection of traumatic memories:

> In their stories of wandering in terror, people uniformly listed the names of every foraged item of food and where it was found: every brook crossed and every hill scaled: every spot where they stopped to rest or eat or sleep. Piling place name upon place name, people's sagas of 1945, as noted earlier, appeared initially to be as irrelevant for Ilongot ethnography as they were impossible to suppress. Misled by the long string of place names, I was slow to realize that what seemed to be at least a year of nothing but hapless wandering actually took only 25 days.
>
> Eventually I came to understand that my mistaken impression was based on half-truth. The recitation of place names in Ilongot stories is designed to invoke the culturally valued notion of movement through space. Beyond their sheer beauty, such lists indicate that a story is true because the teller has actually seen the named places. (Rosaldo 1980, 123)

There is an ethnographic tradition that inspires heritage studies to focus on the study of experiences of contemporary and historical landscapes, considering urban and natural places created by memories and narratives that may still be revived today through the sensibility of participant observation. Through this perspective, processes of destruction and construction can be inseparably attached to narratives of purposeful or accidental abandonment and erasure (Gordillo 2014), complicating the relationship between the destroyed and the factors of destruction as well as bringing into the forefront the narratives that give meaning to heritage value. In a recent post-tsunami account of movement across the Acehnese territory, from Medan to Banda Aceh, the narrators note through their descriptions the way that the "two catastrophes blend into each other and, without changing the structure of that society, change its tonality" (Siegel, Barker, and Djati 2008, 21), but note at the same time that accounts captured in this journey were often without a moral or even chronological position (ibid., 16):

> But in the landscape, where the event is memorialized in the reconstruction, in the rows of identical houses, and in the unreconstructed remains, the event is still present. It is nature turned into buildings. Buildings embodying not the imperatives of right behavior proper to a house, a market, a show, but nature embodied in the fact of destruction and the necessity for replacement. In the identical buildings, where rich and poor are alike, there is a level of differences, an effect of death.

Documenting the efforts that Banda Aceh put toward presenting itself to the rest of Indonesia and the world directed my key avenues of enquiry toward more in-depth ethnographic assessments of the local instruments that give heritage constructs their substance and meaning through the interplay of landscape and narrative. For the period of 2007–2010, I conducted ethnographic research in Aceh and Aceh Besar, focused on the key interlocutors and institutions that partook in the heritage reconstruction efforts and debates, but also invariably capturing a running commentary that sometimes circulated among Acehnese citizens regarding these heritage-related efforts. A significant motivation for the official commemoration of the tsunami was a new vision of the city as a tsunami tourist destination, and the Acehnese themselves could be said to be at the same time commoditized as the objects of disaster tourism, providing a human note to the tragedy, and as tourists, consuming the developments and infrastructure of disaster being offered in their own territory. A 2005 brochure on *Wisata Tsunami* (Tsunami Tourism), released by the regional tourism office, introduces the potential of the tsunami ruin as tourist destination, setting the scene of a landscape in flux through first-person accounts of those displaced by the destruction. The widely circulated leaflet explained in

English that this landscape is composed of places and their narratives, as "the remaining marks of the calamity will pervade the stories of those who directly experienced it and will keep living in the remaining stuffs" [*sic*]. The brochure brings attention to the materiality associated with the earthquake and tsunami: the remaining religious structures, mass burial grounds, and ruins. Moreover, referring to Aceh as a tourist "resort," it claims that this status was "proved . . . when a number of presidents, artists, celebrities and local tourists visited Aceh to have a close look at the marks left by the tsunami," making no distinction between tourists and politically driven photo-opportunists. A 2007 tourist map of the Aceh municipality provides a list of tourist attractions that assemble places of touristic, historic, natural, and commemorative significance. Mass graves shared this cartographic representation with hotels, and the PLTD Apung 1 with the Baiturrahman Mosque, all reduced to artifacts of tourist interest. The persistence of this characterization of the city as it continued to be reproduced through the years reflects the regional government's desire for prosperity gained through economic incentives of local and foreign tourism, an industry that had been unfeasible during the years of separatist conflict.[13] As a milestone of reconstruction was celebrated around the fifth-year anniversary of the tsunami, the city had yet to see any significant increase in tourism. Stopover tourists on the way to diving in Pulau Weh, seeking an exotic destination that is unusually secular, were the predominant type of visitor. Facing what was to many people an unwelcome presence due to their inappropriate dress and motivations, some Acehnese preferred to imagine a tourism exclusively centered around the historic and contemporary significance of Aceh as the *Serambbi Mekka*, an Islamic tourism associated with the *Hajj* pilgrimage,[14] the reiteration of a historic route. Yet the intensification of developments and infrastructure constructed to cater to a loosely defined tourist destination continued in Aceh despite ambiguous and restrictive availability of visas to the region for non-Indonesians. By the time of the fifth-year anniversary in late 2009, and perhaps in preparation for it, this infrastructure was put to work. The residential houses surrounding the tsunami commemoration site of PLTD Apung 1 were turned into souvenir shops, selling mostly refreshments and "I ♥ Aceh" t-shirts, among other goods. To paraphrase one of the t-shirts on sale here, Aceh "never dies," an abstract reference to the separatist conflict as well as the aftermath of the tsunami, a story of survival and resilience that repeats itself. Through this conscious attempt at self-representation, the heritage of Aceh was packaged under three themes that I describe in the following section.

Ruins and *Ruiners*

As time takes its toll with particular emphasis on the Sumatran built and natural landscape, the undeniable effects of oblivion on once-monumental structures of post-tsunami Aceh exemplify a failure of certain monumental constructs to become integrated into the social lives of Aceh. The transformations of the landscape fit not only the imaginary iterations of a new Banda Aceh but also a shift in the authoritative permissions of identity construction. A stark reminder of this was Martinkus's commentary on the abandonment of the "Km 0" monument in Sabang (Chapter 2). As monuments have equal chances to become heritage or ruin, they are subjected to vital reconfigurations one way or another. Stoler (2008) has argued that ruination is a corrosive process that weighs on the future and shapes the present, proposing the observation of sites as condensations of alternative senses of history and ruins not as monuments but as ecologies of remains, open to wider social topographies. The post-tsunami landscape of Aceh is a reminder that ruins are not found—they are made (Abu El-Haj 2001). The epistemic qualities of the process of construction of destruction therefore highlight the significance of considering the process of ruination not as a side effect in the construction of a post-tsunami heritage assemblage but as a central construct in the expressions of destruction and construction as they navigate the underlying politics of forgetting, abandonment, and the lingering pressures of heritage preservation priorities.

There was a type of ruin that marked the territory of the tsunami unconventionally on the urban landscape of Banda Aceh—the tsunami memorial pole. The plaque on one of these *tugu* (monument) in Peulanggahan explained,

> Monument #40. Tsunami. 26 December 2004. Height of the water 4.50 m. Distance from beach 2.50 km. Arrival time of the wave, approximately 8.30am (30 minutes after the earthquake. Magnitude: 8.9 Richter scale).

This simple, concrete structure (Figure 4.1) marked the height, intensity, and timing of the tsunami wave in a particular location, building a permanent display of disaster topography around the city. None of the BRR interviewees had noticed these monuments or heard of this initiative, but there were eighty-five poles assorted across the city as a result of this coherent effort. Hirokazu Iemura, the spokesperson for this project and professor of earthquake engineering at Japan's Kyoto University, was quoted as saying, "Tsunami victims seem to be hoping to forget about the disaster as soon as possible [but] . . . They should learn to cope with the risks in their everyday lives, keeping in mind what happened there" (Iemura and Pradono 2007). The Japanese government spent 9.78 million yen ($85,800 at the time) to finance these memorials, which were approved by the government and the Embassy of Japan in Indonesia. They proposed the construction of one *tugu* for every corner of the "Tsunami-attacked City" (ibid.). Their value, as educational monuments for disaster prevention, revealed itself to be vulnerable from the beginning, as their meaning and significance were lost in the landscape under an overbearing surfeit of

Figure 4.1 Tsunami Memorial Pole, Peulanggahan, Banda Aceh (source: author).

new icons, and they went generally unnoticed. As a result, these monuments joined other NGO ruins left behind with no identified stakeholders or steward. Without a clear role in the reconstructed society, they were vulnerable to being quickly abandoned and falling into disrepair. With the high number of NGOs working in such a short span of time, this process of construction and disrepair was bound to be a significant legacy of the recovery process of Aceh.

After five years of reconstruction work, in 2009, the urban landscape of Aceh had been reconstructed extensively. And alongside the construction of infrastructure, a practice of preservation was observed together with practices of abandonment, justifying certain forms of destruction. Some of these emerging ruins were classic "NGO ruins," the abandoned results of a disconnection between the assessment of needs and the infrastructure provided by reconstruction agencies, caused or exacerbated by bureaucratic obstacles. Abandoned rows of NGO housing lined the west road of the Aceh province from Banda Aceh as far as Meulaboh or even Tapaktuan. Just south of Lhok Nga on this same road, a monumental sculpture peeked above low greenery barely visible from the road. In Banda Aceh, directions to this half-hidden place without name or history were nowhere to be found. But there it was, south of the TNI base, triply obscured behind a row of abandoned shops, a locked gate, and a wall enclosure. The ornate gate at the entrance was designed in the shape of ocean waves. Inside, the paving on the floor appeared to be new but had already begun a losing battle with the rapidly creeping Sumatran vegetation. Then at the far end of the enclosure was the gigantic sculpture, in the form of a three-crested wave, whose tip could be barely seen from the road. The sculpture was hollow and clad in sheets of steel, surrounded by a shallow reflecting pool that was now stagnant and putrefactive (Figure 4.2). This was due, in part, to the Sumatran climate, which sets a unique standard and speed for the processes of decay of material remains. The site represented the perfect symbol of abandonment following purposeful construction and investment in memorialization. Consensus among the different drivers and guides that accompanied me there was that the walls within the enclosures were intended to feature the names of the deceased, but they remained bare. Only six months later, the already dramatic site had undergone an even more dramatic transformation. The metal cladding was gone, leaving only the eerie skeleton of a three-crested wave (Figure 4.3). The vegetation continued to advance, as if there was no doubt that the site was no longer in use. None of the informal interviewees whom I approached in Lhok Nga around that time had ever visited or heard of this memorial, including people who drove their motorcycle to work at the cement factory just south of the monument every day. But they remembered a former school on that same location, wiped out by the tsunami. The present abandonment of this site spoke of a disconnection between the practice of remembering and the practice of memorialization, the latter involving a formality that requires periodic investment in

Figure 4.2 Tsunami memorial in Lhok Nga in early 2010 (source: author).

order to remain active. This example reiterates a complex relationship between memory and place that can only be mediated by time, since the significance that is mobilized for a place to be memorialized is chronologically bound by the context of relevance. Sites such as this one also demonstrated how a cultural heritage battle was being waged silently through the slow recolonization of landscapes in Aceh, many of them previously militarized, with a plethora of tsunami imagery.

True Water

Regardless of any international identity that may have come to be attached to the catastrophe in Aceh within the global mediascape, the tsunami that struck this region is known locally and simply as "the tsunami." In Aceh, some informants were quick to point out that there is a local term used for this natural but rare phenomena, *ie-beuna*, which means "true water." The high death toll then raised new questions about the frequency and magnitude of tsunamis in the past, and the ways in which the memorialization of this event might educate the inhabitants of Aceh regarding the risks of such a phenomenon in order to

Figure 4.3 Tsunami memorial in Lhok Nga in mid-2010 (source: author).

help mitigate future losses. A new vision of a reconstructed Banda Aceh encouraged the permanent incorporation of disaster in the region through memorialization and commodification of this aspect of identity. The production of a post-tsunami disaster identity, particularly in the way that the catastrophe was materialized in Aceh with exaggerated force through the vision of the touristic tsunami city, appeared to intensify over the years. Prolific discussions of disaster tourism and other iterations of this concept (K. Adams 2001; Dann and Seaton 2002; Lennon and Foley 2004; Rojek 1993; Seaton 1996; Tarlow 2005; West 2004; Yuill 2003) have pointed out that a demand for the experience of trauma is not well understood, and therefore, it risks depoliticization. An aspect of this debate that is often overlooked is the role of the articulation of a disaster culture for the production of postdisaster material culture. In Aceh, this absence lends increasing significance to discussing the process of creating a disaster identity in the years following the tsunami. I argue that heritage constructs are instrumental in this articulation, providing an opportunity to recenter accounts on the experience of trauma without necessarily needing to keep coherence with any other aspects of identity. And yet, as local understandings of natural phenomena compete against scientific understandings of causality, the postdisaster identity of Aceh was extremely vulnerable (cf. Blong 1982). This is because local discourses can come under pressure from other structures of cosmological and moral order that have their own ulterior motives, such as religious conversion (Davies 2002). One example of this is the documented adoption of foreign ideas about the apocalypse by indigenous groups, an adoption that has had significant consequences in the shaping of local conceptions of nature and history (Ballard 2000). Recognizing a local disaster identity that is vulnerable to interferences, disaster studies in anthropology have highlighted a cultural construction of reality that shapes risk perception as being critical for future adaptations to the environment (Oliver-Smith and Hoffman 1999). Therefore, a lack of experience—or, more precisely, a lack of transmission of any past experiences through myriad cultural constructs—leaves disaster awareness at risk of ideological contributions from other social, physical, and cosmological settings.

Interpretations of disaster imagery, lost in translation in post-tsunami Banda Aceh, illustrated this vulnerability. A compelling example was the open-air display in the memorial next to the PLTD Apung 1 tsunami ship that includes photographs of the catastrophe. These already ageing photos at the time, printed on A4 paper and without any protection from the elements, included images of tsunami victims and debris, displaying dislocated people and urban landscapes alike. At the center of the collage, there was a photograph that stood out: a printout of a satellite photograph of what at first appears to be a cloud in the shape of a spiral. Upon closer inspection, one could identify this image as a view from space of an atmospheric phenomenon. The coastline of the peninsula of Florida in the United States could be made out, alongside nearby Caribbean islands. Given the chronological proximity of the Hurricane Katrina catastrophe that took place soon after the Indian Ocean tsunami, in August 2005, it was natural to assume that this photo was the product of a misclassification of images on the World Wide Web. At the time, it made sense that the imagery of two globally significant disasters had been accidentally conflated, or perhaps, an expression of solidarity toward "the next disaster" was being put on display intentionally. The image—right in the middle of the panel—suggested an intentional and meaningful inclusion in this visual assemblage, but left without a caption it was particularly vulnerable to misinterpretation. Opportune interviews with visitors at the memorial, who identified as Acehnese or Indonesians from other provinces, revealed different but elaborate interpretations, with some shared underlying themes. One of the guides who made himself available at PLTD Apung 1 explained that the image depicted "the hole that the tsunami created in the ocean." This theme was overwhelmingly shared among most visitors, who tried to make sense of the assemblage of images as a whole rather than the individual image in question. Where I saw the outline of the coast of Florida, many saw the coastline of Sri Lanka. Where

I saw clouds, many saw waves. Where I saw the eye of a hurricane, they saw a sign. The idea of a hole in the ocean should be seen as related to the violence with which the tsunami was experienced and the assumption that it must have been a violent event visible in the ocean as well. Moreover, it occurred to me much later that the way that the tsunami was predominantly depicted cartographically—as a concentric design with a center at the epicenter of the earthquake and circles representing the wave expansion across all directions and reaching the mainland—could have significantly influenced the way that it was imagined to have taken place in real space and time. The fact that the tsunami forces were for the most part invisible as they travelled across the ocean and could be noticed only as they reached the coast was not considered in this narrative. Later, I discovered that the satellite-generated image in question had been circulated for some time,[15] wrongly attributed to Hurricane Frances in September 2004, to the Indian Ocean tsunami in December 2004, and to Hurricane Katrina in August 2005. The event photographed was, in fact, a depiction of the approach of Hurricane Floyd in 1999, as documented in a NASA database. Even the circulation of scientific images is vulnerable in a mediated world where fact, factoid, and imagination are archived together, easily dissociated from the source that gives each type of knowledge legitimacy.

What contributed significantly to the perseverance of this type of myth is that the force of the tsunami had hardly been legitimized outside of the scientific discourse. A recent Indonesian book has collected a series of conspiracy theories surrounding the catastrophe, associating the disaster with events as unlikely as a nuclear explosion in the ocean by the United States or perhaps France—both explanations suggesting an awareness of the historical impact that these two nations have had in the ocean through the controversy of nuclear testing (Dzikron 2009). The darkened, putrefying appearance of tsunami victims floating on the Kreung Aceh River and washed ashore on the coasts was interpreted to be proof of charred human remains due to discoloration. Another potential anthropogenic cause of the disaster that became a suitable explanation to many was the endemic separatist conflict of the region. A dominant discourse of post-tsunami Aceh promoted the tsunami as a pacifying force of change, perhaps a last resort. Proposed as divine punishment for the region's bloodshed, the causality of the tragedy confirmed the agency of Allah as it was perceived to operate in this disaster. Destruction by an almighty force was unrelenting but compassionate enough to offer an opportunity for peaceful rebirth, a blessing in disguise to help the Acehnese get back on the right path of peace and other *halal* Muslim practices. Material evidence to support this perception was compelling: the fact that many mosques not only survived the ravaging effects of the earthquake and the ensuing tsunami waves but also provided safe refuge for many survivors was often noted. The images of a crumbled landscape where debris surrounded the relatively intact

Baiturrahman Grand Mosque in Banda Aceh, the Baiturrahman Mosque in Ulee Lheue, and the Rahmatullah Mosque in Lampuuk were quickly interpreted as material witnesses to the intentions of an omnipotent force. The skeptics simply attributed this fact to differences in the quality of structures and construction between residential and religious structures, a likely conclusion in corruption-ridden Indonesia. But this visual reminder of the will of Allah did not stand in isolation. Many Indonesians explained that the timing of the tsunami was not accidental, an explanation that was key to understanding the construction of causality and the subsequent place that the disaster would take in collective memory. December 26, the day after Christmas, was generally a tense time around Aceh. It was believed by some that an inappropriately emphasized celebration of the Christian holiday in Aceh this time around had caused spiritual tensions in the most Islamic of Indonesian regions.

Islamization of Catastrophe

A divine agency was expressed in a different interpretation of the satellite photograph shown at the PLTD Apung 1 memorial. A visitor confidently explained, while gazing at the image, "This is probably a calligraphic representation of God," and elaborated, "and this is the spiral shape, just like the way we walk about the Ka'ba during the *Hajj*" (pilgrimage). It was an image that was echoed in the design of the museum: the symbolic eye of God and the spiraling path toward salvation that evoked circumambulation. As the fifth-year anniversary approached, banners lining the streets and buildings of Aceh reinforced this belief, associating penitence and catastrophe with the season for commemoration. One of them read: "Make the tsunami of 26 December 2004 as a reminder for us to repent to Allah PBWH"; "Commemorate the 5 years of the earthquake and tsunami with prayers and remembrance to the great powerful God who protects." Interpretations of the tsunami imagery singled out the territory of Aceh as intrinsically Islamic, and increasingly spoke of a disaster that was God-driven rather than associated with inherent vulnerabilities and manufactured risks.

A sign that confirmed that the tsunami boat sites had formalized their place in Aceh as sites of memory was the collection boxes that became prolific in each place. At the commemoration site of the Lampulo tsunami boat, a red box explained to visitors:

> Charity Box
>
> 1) to assist (alms) money to the construction of the mosque of Lampulo village.
>
> 2) God will recompense
>
> 3) Please understand [fragmented].
>
> Signed, the mosque committee

Similarly, at the site of the PLTD Apung 1 boat, a pink box awaited donations with a sign that explained, "Tsunami stands for: T: *tuhan*, S: *suruh*, U: *umat*, N: *nabi*, A: *agar*, M: *manusia*, I: *Insaf*." This was consistently interpreted by Indonesian visitors to mean, "God told the prophet's followers to become converted."

The tsunami boats became the perfect candidates for containing and transmitting messages regarding the spiritual significance of the tsunami. During their unintentional relocation to their respective resting places inland, both tsunami boats were said to have "rescued" a number of people, a lifesaving role that was fresh in the collective memory of these unlikely monuments. Moreover, as the collective narrative went, the larger boat managed to avoid breaking any mosques on its forced path, and there were many. Hence, the boats were not simply left in place because of a logistical inability to move them back to the water, a serious consideration for the larger ship. Rather, they were discussed as a representation of *ayah*. *Ayah*, the verse in the Quran about revelation and proof, is a way of seeing and reading the world around oneself to find the message of God. It is a message that comes from things, not just verses. Accordingly, on tsunami commemoration days both boat sites were completely cordoned off and inaccessible to the public while prayers were taking place. It was then clear that the tsunami boats had become effectively Islamicized though discourses and practices. The boats could be the *ayah* sent through the tsunami.

In a region where no more than 4 percent of the population adheres to a religion other than Islam, the legitimate religion of Aceh can be considered to be Islam. Other religions are associated with migrant communities from other regions of the world and therefore do not compete with Islam on the grounds of indigeneity. Clifford Geertz (1968) argued that Islam in Indonesia may operate as a thin layer brought by mobile communities from Arabia and applied over an indigenous animism. This appropriation, he argued, results in a lack of uniformity of Islamic practices (ibid., 11–12) and a "struggle for the real" (ibid., 104) between secular and religious interpretations of the world. Geertz described this struggle in the region as chronic and intense, rendering the history of Islam unintelligible. Yet, the city's identification as the Verandah of Mecca symbolized an affirmation that Islam came from somewhere else without affecting its unwavering appropriation of the Mohammedan belief. Therefore, the heritage associated with the Verandah of Mecca, as well as the heritage instruments that give it meaning, were increasingly meaningful to the reconstruction of the city. The calls for reconstruction were seized by religious revivalists in Aceh (Feener 2014; Jauhola 2013), who used this strategic moment to propose further transformations in a society strengthened by the implementation of Shari'a law. Marjaana Jauhola describes how the discourse of BRR later emphasized the prominent place that religious aspects had in the rehabilitation and reconstruction efforts, claiming "the risks were too great for

BRR to focus merely on rebuilding damaged houses and infrastructure" (BRR 2009, quoted in Jauhola 2013, 4).

In a recent study of cultural factors involved in the process of reconstruction of Aceh, Patrick Daly and Yenny Rahmayati (2012) suggested that the mobilization of a perception of tradition and its disappearance was responsible for a failure in the construction or enhancement of resilience. The reconstruction policy of BRR assumed that adaptation was adequately represented in heritage constructs, attempting to reconcile past know-how with future resilience in the urban fabric of the region. But not all reconstruction was being carried out in accordance with the needs of traditional livelihoods. The early days of reconstruction saw a construction frenzy of housing and other infrastructure carried out by a variety of NGO initiatives. With the *rumoh Aceh* as an iconic heritage image, some Acehnese critiqued the design of many of these efforts as deviating from the norm (e.g., an elimination of verandahs). But a more significant faux pas was the inability of some of these initiatives to identify the different purposes and social roles of particular structures. Although architecturally they could be perceived to be similar, the *mushollah* (small mosque) and the *meunasah* (the traditional center of community activities) serve notably distinct purposes. The *meunasah* was absent in many reconstruction plans, a failure to incorporate necessary social and communal roles in urban reconstruction planning. These spaces are key for the performance of heritage as the authorized place of legalization of village decisions, concerned with building character and regulating behavior. This was also evidence of restrictions internal to the work of many NGOs—whose work was to remain secular—toward the construction of religious structures. As a result, the new layout of Aceh was disturbing to many, and destroyed traditional livelihoods and the significance and use of traditional forms in ways unexpected by many, and outside of typical concerns with a "heritage at risk."

Time and Timeliness

With BRR already gone, the fifth commemoration of the tsunami on December 2009 represented an artificial end to an ongoing rehabilitation process. The grand finale was set to take place on the fifth-year anniversary of the tsunami, when the city of Banda Aceh would be celebrated as a renewed city, as it had been ideally imagined—Tsunami Day. For years, the center of attention in Banda Aceh had been the offices of BRR, affectionately called "Biru Rapat-Rapat" (Bureau of Meetings, in Acehnese) by some of its Acehnese employees who were interviewed. Its centrality for the early years of reconstruction was undeniable: "A fulfillment of the people of Aceh and Nias which is dignified, trustworthy, democratic and prosperous—BRR," as the plaque in the center of the compound read. But with BRR dissolved, the center of official attention on Tsunami Day shifted to Ulee Lheue, the revived port complex in the Pante Cermin village, often called "ground zero." The official ceremonies were planned from 9 a.m. to 12:30 p.m. on December 26, 2009. A leaflet, available at the regional tourist office, explained the series of events organized and hosted by the Department of Culture and Tourism of Aceh, in conjunction with the *Badan Kasinambungan Rekonstruksi Aceh* (Agency for Sustainable Reconstruction, BKRA) and the Multi Donor Fund (MDF). Attendance at the event was expected to be on the order of six thousand people. But the event at Ulee Lheue turned out to be more exclusive than imagined. The gates of the port were closed, leaving some Acehnese and many *becak* (motorcycle-taxi) drivers waiting outside to see what was happening, together with any members of the press who did not hold the correct press badge, as a reporter from the Associated Press explained, joining us outside the gates. Aside from the vast number of Indonesian flags lining Ulee Lheue's walls and substantial armed forces guarding the fence, the most noticeable snapshot of the morning was the presence of a battleship patrolling the coast. Throughout the week, the presence of Brimov (mobile police brigades) and TNI operatives was undeniable. Guarding the inside and surroundings of the Aceh Tsunami Museum, their presence included a black armored vehicle pointing a gun at the museum door from the park across the street. Many sights were closed off in preparation for the official visit of the vice president. In addition, it was the day after Christmas, and I was told that Banda Aceh was unusually militarized around this time in order to mitigate any potential riots associated with Christmas time in this most Islamic of Indonesian territories. On Christmas Day in 2000, which coincided with Idul Fitri, thirty-eight churches in eleven Indonesian cities were reported to have been sent bombs (Martinkus 2004).

None of my close Acehnese contacts and acquaintances attended the commemoration events on this day. Perhaps the high density of Brimov and TNI forces on the streets served as a deterrent. The remembrance for them was happening instead in their homes, in private ways. Stories were dusted off to tell the guests: "Oh, you went to visit PLTD Apung? I have a friend who died there," said my friend's father. And then he added, "Well, it *was* the tsunami!" as though to remind us that when the topic of the tsunami is discussed the story is truly and expectedly about death. Everybody had a tsunami story. And for another friend in Lhok Nga the stories of tsunami casualties eventually mingled with other family losses. Dani told me earlier that week, while we were discussing the catastrophe, that he had a younger brother who had passed away. He said, "He was walking on the street when a TNI soldier shot him right there in that *kampung*." We did not continue in the direction he pointed. Tsunami victims could be added to an earlier body count, reported to be up to fifteen thousand people, after decades of separatist conflict. Clashes often caught civilians in the crossfire, regardless of whether they took sides. For many Acehnese, the landscape of terror left in the wake of the tsunami overlapped with a previous landscape of trauma.

Unable to witness the celebrations at Ulee Lehue, we sipped coffee outside and studied the tourist leaflet instead. A call for reflection on the five years after the tsunami drew attention to the tsunami as a stepping-stone for peace-building efforts, adding a positive note to the loss of hundreds of thousands of lives and refocusing the disaster on the construction of peace rather than the destruction of livelihoods. The plan for the commemoration included a prayer and recitation in meaningful locations as well as an opportunity for the BKRA to host a seminar on "Aceh's Development Prospects Post-Tsunami." The series of official "field trips" scheduled in the afternoon matched the heavily guarded locations that had been observed earlier in the day: the Ulee Lheue mass grave, the Aceh Tsunami Museum, and the Grand Mosque Baiturrahman. An otherwise normal day, the "tsunami heritage" was the center of this performance. And then the Aceh Tsunami Museum finally opened its doors to the public. On Tsunami Day, the museum remained closed through the morning, a Brimov tank standing guard on the street. Meanwhile the local and international media feasted on a courtyard display, by the International Red Cross, of Acehnese children holding umbrellas that depicted an assortment of children's faces, perhaps intended to represent through their different skin colors a melting pot of cultures. The institutional scene was transformed in the afternoon when thousands of people flooded the garden surrounding the structure. Street vendors occupied the courtyard and lawn, selling food, drink, toys, and even decorative goldfish. A drum ensemble attracted a cheering mob. Playing nonstop in the shade of the open ground level of the museum, they held forth by the reflecting pool that had always been the busiest part of the museum structure. Within the museum, glossy posters offered an abstract, translated *bahasa arsitectur* (architectural language), with areas named evocatively and linearly, from Spaces of Memory, to Spaces of Hope, and finally Spaces of Relief. The structure asked for an embodied experience of Tsunami Day: a dark and narrow corridor lined by two floor-to-ceiling waterfalls—the Tsunami Passage—and the ambient sound of thunder aimed to "remind us of just how scary the atmosphere was." Through the Memorial Hall, a small dark auditorium, came the Blessing Chamber, a dark tubular space with a skylight depicting a calligraphic representation of Allah, the Light of God, which "shines light as a symbol of man's relationship with his Lord, its presence and hope for the community of Aceh." A Space of Hope followed, in the shape of the spiral ramp that ascended toward the open atrium of the museum's Atrium of Hope, crossing over the reflecting pool. From this open space, one could appreciate the natural hill on which the museum was built, the Hill of Light that remembers those who approached it for safety. Whether the hill was actually used to run for safety was, like many tsunami stories in Aceh, partly legend, but it had nonetheless inspired the name of the winning design for this museum: "*Rumoh Aceh*—Escape Hill."

In this way, the heritage of disaster for Aceh was created through discrete moments, such as the ones described earlier, rooted in the *now*. Able to be detached from the exclusive burden of history, ethnographic perspectives on the construction of heritage status suggest that the range of factors that need to be examined in order to construct sincere heritage perspectives are infinite, creative, and challenging. As part of this undertaking, what I also aim to emphasize in this chapter is the volatility of certain states in preservation that are taken for granted, such as the state of being "under preservation," "under construction," and "under destruction," states of conservation that are in fact able to coexist at different moments or concurrently from different vantage points. These chronologically and spatially localized descriptions also demonstrate that heritage constructs can effectively mediate and resist the victimization of culture, taking an active formative role in the construction of futures that are locally sketched to transmit disaster narratives hereafter. Incidentally, these heritages enrich the resilience capability of Banda Aceh in accordance with existing cultural and religious frameworks that give local meanings to disaster and destruction.

5 Destruction Alternatives

The heritage of that which has been destroyed—the heritage of disaster—is not invisible. Independently of its manifestation in material culture, it continues to exist in discourses, ruins, and reincarnations in new forms of heritage, such as the ones discussed in this book. The heritage of disaster is a material culture constructed on interpretations of permanence that are flexible and dialogical, accommodating widely discrepant meanings, and drawing on the temporal flexibility of a product that is contemporarily shaped from history (Jackson 2008; Turnbridge and Ashworth 1996, 20). In a temporal construction that considers above all its presentness, heritage is "heritagised" (Harvey 2001, 3), defined in *the now.* However, while it gives time a privileged position in its process of construction of value (the older, the more valuable), it is also ironically characterized by a decontextualization of chronological reality in which heritage *is*, regardless of the *when.* This clashes the most, I have argued throughout this book, in the mastering of destruction and articulations of risk, which hinge on artificially bounded moments. Engaging with such an epistemological problem, this concluding chapter discusses the significance of addressing alterity in critical heritage studies with a recognition of destruction as a legitimate constructive practice at its core.

Reclaiming *Post-Heritage*

In the last scene of Gabriel García Márquez's *One Hundred Years of Solitude*, while the fantastic story of generations of the Buendía family and their fictional town of Macondo comes to an abrupt but predicted end, we are left to wonder whether the story was a chronicle of a family or the life history of a place. In the end, the heritages of kinship and town are intertwined in such a way that they are both reduced to nothing, erased without legacy. This book does not aim to trivialize erasure, but rather, inspired by this complex literary ending, to bring agency to destruction by "peopling" the heritage landscape. In doing so, it aims to reset perceptions of cultural heritage to accommodate destruction more constructively through a wider appreciation of the way that heritage is subjected to *change*—a force that makes heritage visible or invisible, bygone or prospective. Here, I am influenced by recent scholarship on absence that acknowledges the material affect of gaps, voids, traces, and ruins in the material, social, and political landscape (e.g., see Bille, Hastrup, and Sørensen 2010; Buchli 2015; Buchli and Lucas 2001; DeSilvey and Edensor 2013; Fowles and Heupel 2013; Gordillo 2014; Harrison 2013; Herscher 2010; Hetherington 2004; Meskell 2002; Meyer 2012; Olsen and Pétursdóttir 2014; Stoler 2013).

There is an apparently unremarkable intersection in Lhok Nga next to the tsunami mass grave. Although it looks like any winding street of Greater Aceh, there is a triangular island of grass at this intersection, featuring a relatively new concrete monolith. Here, a plaque provides a caption for the surrounding landscape, describing the existence of a former airstrip on the current location of a golf course. Maimun Saleh Air Field, the name of the airstrip, pays tribute to an Acehnese pilot who flew for Indonesia's Air Force Fourth Squadron in the 1950s. The plaque explains that he died in 1952 and was memorialized on April 9, 1954, by the Indonesian Air Force with the naming of the military airfield at Lhok Nga. He was the first Acehnese pilot. My informant explained that whoever built the monolith hoped people would think, "Oh, Maimun Saleh is mentioned there, and what happened there? Oh, and who is Maimun Saleh?" So the commemorative plaque was created to answer questions that had not yet been asked, about a destroyed heritage landscape open for reinterpretation. There was a hope in this vision of commemoration that the resurrection of an invisible place would add meaning to this forgotten landscape. Today, this relatively unknown snapshot from the past survives though this plaque, mostly witnessed by speeding motorbikes. Nonetheless, the ghost of the Maimun Saleh airport[16] is now formally part of an assemblage of heritage places, tangible and intangible, that was imagined and formalized after the vast destruction of the tsunami. This particular site joins the collection of invisible heritage that survives its own destruction, brought back to the surface through trilingual plaques throughout Aceh and Greater Aceh, such as the ones already described, and their inclusion in a series of heritage trails. Despite the timing of this assemblage, most of these sites bear no association with the tsunami catastrophe, but the opportunity for redescriptions of this landscape converges nonetheless on the aftermath of destruction in the 2004 tsunami. Again, the significant erasure of the landscape attributed to the 2004 tsunami presented an opportunity to make space not only for the emergence of new heritage forms but also for the reorganization of the existing heritage assemblage in order to reflect new priorities and emphases in the chronology of Aceh.

Because I was standing on an airfield, I was reminded of the term "airmindedness," made popular in the 1930s to refer to a state of enthusiasm about flight as a new technology.[17] It entails a set of cultural traditions, symbols, and markers that, combined with the existing political culture and social institutions, constituted a given nation's response to the airplane (Palmer 2006). Specific manifestations of airmindedness were defined as the product of each nation's unique historical and cultural traditions, as they revealed the entanglement of a nation's historical narratives with technological solutions to an ever-expanding world. To be airminded was to believe unquestionably in the use of aviation as a tool for shaping or relating to the physical world. Once air travel was democratized and eventually taken for granted, airmindedness was achieved, and the word became redundant (Edmonds 1993). As Leigh Edmonds states, the era of aerial evangelism was over. There are many aspects of airmindedness that could be transferred to the process of transformation of any landscape into a landscape of cultural heritage—a process of engaging with heritage technologies through "heritagemindedness," the popularization of heritage as a primary vehicle and tool for experiencing, shaping, and interpreting the world around us. Likewise, it could be said that specific manifestations of heritagemindedness are the product of each nation's unique historical and cultural traditions, a localized "process of engagement" (L. Smith 2006, 1). Heritagemindedness is ruled by standardized approaches and an overwhelmingly powerful rhetorical language that define what is, and what is not, heritage (Lafrenz-Samuels and Rico 2015). The construction of a heritage landscape in Aceh challenges many of these foundational elements, suggesting that the example of Aceh alters the trajectory and legacy of heritagemindedness in "alternative" contexts. The account provided here suggests a number of ways in which heritage might be reimagined and done differently, in ways that open up a more democratic and vernacular engagement with heritage as a creative, regenerative, and future-oriented pursuit. A different type of heritage demands a different approach to its understanding.

Situating Vernacular Subjects

Ethnographic heritage fieldwork faces an insurmountable problem: working on a research subject that is at once constructed using situated value-laden concepts, a product of localized and temporalized interactions, and which exists simultaneously in conversation with regional, national, and international agendas and narratives that must be translated and fit into a coherent modern and cosmopolitan construct. But the use of "heritage" as a unit of discussion in fieldwork has a noted tendency to steer conversations heavily toward familiar images of monumental sites and established standards of practice. This is part of the rhetorical instrumentation that can be said to have now reached all localities, establishing a moral and natural order (Foucault 1970) in its classification that is empowered by the dominant and self-referential discourse, a heritage construct that becomes heritage "*because* it is subjected to the practices of management and preservation, not because it simply *is*" (L. Smith 2006, 3, her emphasis). Heritage as a category nonetheless could be said to be unreal (G. Lakoff 1990), but an abstract entity where no member of the category should be more representative of the category than others, since the category is defined only by its properties. Therefore, to problematize the adoption of an externally imposed category such as "heritage" in the search for indigenous heritage expressions is to problematize the very construction of the category of heritage itself. More than this, to seek a heritage expression that is not articulated as "heritage" is to ask informants to help construct and legitimize an *absence* that opens itself to be colonized by foreign constructs. This is why, throughout this book, I have highlighted the importance of taking a view of non-Western heritage within the framework of vernacularization (Engle Merry 2006, 39), considering transnational sources that travel to small communities, where they are adapted to local institutions and meanings. This framework, discussed by Engle Merry, asks that attention be paid to knowledge brokers, those intermediary individuals and organizations that channel the information and play a role at adapting global concepts to local conditions, working on local, regional, national, and global systems of meaning. It requires the interrogation of the heritage landscape through the visions and concerns of local and global "experts" alike and a recognition of the way certain technologies are transported through these networks, such as the instrument of risk assessment. In Indonesia, the local appropriation of heritage as a concept encompasses local interpretations of materiality, memory, and identity that interact with globalized portrayals to create a very specific product. The examination of the aftermath of the tsunami catastrophe in Banda Aceh, is concerned with the traditions that negotiate this process, focusing on the intersection between existing cultural traditions of Banda Aceh and the global tradition of preservation—a philosophy and practice that are primarily concerned with interrupting the destruction of material culture, whether physically or discursively. This has been examined through the circulation of the construct "heritage at risk," a rhetorical device that regulates the deployment of preservation at the global level.

However, the uncritical deployment of negative attitudes to destruction to justify and motivate heritage preservation agendas has been shown to have obscured the true construction of heritage value in the reconstructed city. In particular, this is due to a dominant heritage framework's inability to acknowledge the emergence of *new* heritage places and their narrative, as was observed in Aceh following the 2004 Indian Ocean tsunami. This contrast demanded a reevaluation of the ways that vulnerability and heritage have been blended and naturalized in discourse and practice. In the context of an Acehnese claim to uniqueness, the successful vernacularization of a preservation practice is evident in the formalization of distinct ways to represent history, identity, and a recent assimilation of a "culture of disaster" in coherent heritage assemblages. However, the heritage construct that is articulated in Aceh does not fit the already established boundaries of prepackaged "non-Western" or "Asian" heritage debates. Instead, it creatively weaves elements of prearticulated global preservation philosophies to preserve specific local agendas in the designation and articulation of significance. As such, the heritage of Aceh does not require legitimization by the global heritage discourse or its experts, as it responds to localized structures of legitimacy and democratization.

Three themes have been shown to intersect throughout the featured discussions of unique heritage constructions and destructions in Aceh, highlighting the significance of vernacularization. Firstly, the construct of "heritage at risk," as a point of departure in this critical heritage reassessment, is an effective instrument that universalizes the cultural heritage preservation mission, defining a framework that assimilates heritage only through its vulnerability and destruction while remaining vulnerable itself to politicization. Risk also privileges the expert voice and de-privileges non-Western appreciations of destruction as a legitimate constructive process. It marginalizes other interpretations of time, permanence, and even divine agency in the construction and destruction of materiality. "Heritage at risk" is a significant factor that impedes vernacularization of the heritage construct. I have suggested that one way of responding to this is a de-privileging of risk itself, in a new framework that is sensitive to the ways in which heritage might be interpreted to be at risk, and *not* at risk at the same time by different interlocutors.

A second theme is the naturalization of a cultural heritage preservation methodology that takes no account of the hybridization of imported concepts that are merged with local ones in Aceh, leading to misinterpretation and false value designation. As the heritage construct resists integration into either the Western dominant discourse or an Orientalist vision of "Asian heritage," it problematizes the operation of a dichotomy promoted by Western discourses in order to legitimize attitudes to "alternative" heritage experiences in the imagination of global preservation efforts. However, I argue that an examination of the identities and agendas of knowledge brokers in the reconstruction of post-tsunami Aceh remains incomplete and therefore the categorization of local and foreign expertise remains misrepresented. The localization of an Acehnese heritage discourse is a multifaceted challenge, stretching a territorial distance that reaches as far as Middle Eastern heritage discourses in its search for spiritual and historical legitimacy.

The third and most striking theme proposed in this volume is the role that cultural heritage has in the manifestation and transmission of a recent history, in particular, the validation of an emerging "culture of disaster." Locally emergent heritage constructs are rooted in an intersection of sociocultural, historical, political, and religious agendas and guidelines for legitimization. As the heritage of Aceh emerges independently of the constraints of vulnerability and ageism imposed by external heritage discourses, it embodies a tsunami identity that has the potential to play an important role in the local manifestation of trauma, resilience, mitigation, and recovery. Heritage can be lifesaving.

A key subject for interrogation in the study of vernacularization is the individuals and institutions that have the ability to shape heritage constructs. Channels of authority and legitimacy in cultural heritage management have been challenged only from a theoretical standpoint, but remain mostly unexplored from reflexive "internal" institutional positions. This critical distinction is reflected in the way that the two realms of knowledge continue to frame heritage questions through the dichotomy of expert and nonexpert positions (the ethnographically informed viewpoint remains, conveniently, a nonexpert role in many global heritage institutions). There are, therefore, two different ambitious outlooks for the way that cultural participation is envisioned in the work of heritage, which I discuss in the following section, marking a difference in the study of alterity in cultural heritage according to different ideas about the transfer of authority in practice (Wharton 2005).

Heritage Alterity: Theory versus Practice

On the one hand, critical heritage theory has argued that we need to see heritage studies as a situated knowledge (Haraway 1991) that opens a space of communication between different ways of producing said knowledge. Speaking against a universalizing approach to alternative heritages, Anna Karlström (2009, 178) points out that it is critical to recognize a distinction between different *referents*, which is an epistemological question, and different *things*, which is an ontological claim (Karlström 2009, 178). Such recognition acknowledges the existence of different worldviews and of different worlds. This difference has been partially accommodated in cultural heritage discourse and practice under the classification of "alternative," "non-Western," and even "Asian" heritage constructs and approaches, an inclusion that has satisfied to some extent the calls for a decolonization of the discipline, which nonetheless remains attached to dominant forms of expertise and methods of identification.

Calls for dialogical models of heritage management (e.g., Harrison 2013) realistically acknowledge that not everyone has equal access and power to shape heritage discourses and practices. Inherent to the field and debates in heritage studies are firm dichotomies of "us" and "them": a separation between knowledge producers and knowledge consumers, a separation between the public and the state, and a separation between experts and nonexperts—dichotomies that define a tension between knowledge production and consumption that has dominated critical heritage debates since the end of the twentieth century (Carman and Sørensen 2012, 15). This tension has also been discussed as a struggle that emerges between authority (the academic viewpoint) and authenticity (the community experience) in the battle for the control and promotion of values in the public sphere (Shackel 2009, ix). An acknowledgment of indigenous values and practices in heritage has been made explicit in different ways, firmly located in an awareness of "the other," its values and rights with regards to heritage value. The idea of "alternative" heritage value, however, promotes a rhetoric that suggests a hierarchy of mainstream and deviant heritage constructs, and moreover, a rhetorical destruction of heterogeneity within regional or perceptual categories of heritage that are perceived in this hierarchy to be homogeneous. The effective recognition of "local experts" (i.e., agents who transcend and therefore merge the traditional categories of foreign expert and local voices and authorities) helps blur and problematize these distinctions in productive ways, as suggested in this volume.

Participation has been a strong line of enquiry in the study of cosmopolitanism and ethics in heritage, an interrogation that recognizes the need for autonomy while aiming to avoid the burdens of diversity that may entrap people within expectations or traditions of difference (Appiah 2006). Repositioning heritage analysis within a "rooted cosmopolitanism" (Cohen 1992), as it has been suggested in this framework, would bring together multiple spheres of legitimacy on a common ground. A cosmopolitan framework in the study of the construction of heritage may not necessarily be preservationist in ethos nor will it attempt to coerce people into some preserved ancient authenticity (Meskell 2009). It demands, rather, that frameworks for heritage management be rooted in local settings and practices, recognizing difference productively. The view is also wary of becoming a revamped dominant discourse, questioning, for example, the applicability of participatory approaches in nonparticipatory cultural contexts (Rico 2006), and the applicability of postcolonial heritage debates in regions with different colonial histories (Exell and Rico 2014). However, despite compelling arguments to reimagine authority in the curation and preservation of heritage, the standardization of heritage assessment methodologies appears to persevere, suggesting a significant disconnection between theoretical and academic debates that aim to create a space for alterity and the practices of heritage identification, documentation, and management that produce specific forms of heritage construct.

On the other hand, alterity in heritage practice lies in the recognition of multiple voices and agendas, and is identified in heritage methodologies through the inclusion of stakeholders. Stakeholder consultation alongside a development of heritage value typologies has been proposed for the development of a lingua franca that enables the ranking of often conflicting values (Mason and Avrami 2002). This model recognizes the challenge of identifying and legitimizing stakeholders; however, it has been less effective at problematising the harmony with which a deliberative democratic process is imagined to take place. In particular, as it is proposed that "the views of experts, citizens, communities, governments, and other stakeholders can be voiced and compared more effectively" (ibid., 9), this model retains the already dominant role of the heritage expert as a stakeholder, one that is also responsible for the hierarchization of all values. In order to challenge the tendency to engage in ideology transfer inherent in heritage work (cf. Byrne 1991), decentralized approaches to heritage assessment and management have convincingly mobilized the loaded role of expertise in dialectic directions. Involvements with the stakeholder are increasingly recognized as educational encounters for both parties involved, making the modes of transmission of knowledge through institutional and academic channels come under more scrutiny. But failing to make this a prominent requirement in heritage methodologies, the process of stakeholder legitimization remains a part of cultural heritage management that operates within a methodological framework ruled by improvisation and informality (cf. Appadurai 1996). The work of archaeological heritage has proposed to address this challenge through a cocreation of research questions and methods (Damm 2005, 80–84), which are dialogical models of negotiation (Zimmerman 1997, 55) that aim to circumvent the colonial legacy of the discipline (Colwell-Chanthaphonh and Ferguson 2006, 159) and increasingly tested as a way to create the spaces of dialogue that are needed to engage with marginalized stakeholders on their terms (i.e., Breglia 2006; Castañeda and Matthews 2008; Hancock 2008; Meskell 2011). I argue elsewhere (Rico, forthcoming) that a turn toward more technologically advanced approaches to heritage preservation and management (which is particularly visible when it comes to assessments of risk and endangerment) has hinged on the promise of a paradigm change for expert-stakeholder relationships through the application of more inclusionary models of assessment, valuation, and dissemination. However, less attention has been paid to the nature, needs, and limitations of these approaches in practice, and as a result, this can marginalize certain users—the nonexperts and local experts—in favor of promoting a new form of expertise, the technologist. This relatively new expert in heritage might override the authority of local actors and expertise like any other connoisseur embracing the dominant heritage discourse.

But other heritage practices have only recently begun to acknowledge the existence of an "unofficial heritage," a broad range of practices that are represented using the language of heritage, but are not recognized by official forms of legislation (Harrison 2013, 15; Harrison and Schofield 2010, 8–9), evidencing the need for ethnographic approaches in the identification and documentation of heritage constructs. There are a few suggestions that an emerging ethnographic approach to heritage makes moving forward. At a general level, it recommends that professionals be sensitive to or aware of different worldviews (Moses 1992, 3), in order to mediate the identification and translation of concepts in preservation and management that have been created and made operational elsewhere. At a practical level, it raises awareness of the powerful role that a constructed endangerment "sensibility" has in the identification and management of official heritage (Harrison 2013; 2016; Vidal and Dias 2016) and the challenge of identifying a similar relationship between heritage and permanence in other heritage constructs. Practices of destruction are both *meaningful* and, counterintuitively, *constructive*. Destruction has informed instrumental incongruences within the dominant heritage discourse and brought to light some of the most challenging questions in heritage debates (e.g., Layton, Stone, and Thomas 2001). Understanding destruction in constructive ways shows us how we might move away from a vulnerability framework in such a way that a "heritage of change" can be legitimized through traditions in flux. To hybridize the words of one of the earliest dedicated works in conservation on heritage at risk (B. Jones 1986, 11) and more recent debates on the endangerment of digital heritage, the "youngest" and most recognized form of contemporary heritage construct (Cameron 2008, 175), "it seems remarkable that anything at all has survived from the past" . . . a resource "born disappearing."

itself—its nonelite [illegible] and local voices—in favor of promoting a new form of [illegible], the national past. This relatively new [illegible] in heritage [illegible] override [illegible] local [illegible], and generate [illegible] over other [illegible] emphasize the dominant heritage discourse.

[illegible]

Epilogue: "Then and Now"

> What do the remaining marks say to us? History is a valuable experience for human beings. Looking back at the events, it is expected that human being [*sic*] can learn from the past so that such disastrous impacts can be minimized in the future and people can grow wiser. (2005 Tsunami Wisata leaflet, Banda Aceh)

There is a lot at stake in the recognition and management of a heritage of disaster in Banda Aceh that transcends Aceh, Indonesia, and the particular catastrophe of late 2004. During the completion of this project, which focused exclusively on the aftermath of the 2004 tsunami, a number of natural disasters took place around the world: the 2005 Hurricane Katrina in the United States, the 2005 Kashmir earthquake in Pakistan, the 2006 Java earthquake in Indonesia, the 2008 Sichuan earthquake in China, the 2008 Chaitén volcanic eruption in Chile, the 2010 Haiti earthquake, the 2010 earthquake and tsunami in Chile, the 2010 Pakistan floods, the 2010 Penang tsunami in Indonesia, the 2010 eruptions of Mount Merapi, the 2011 Tohoku earthquake and tsunami in Japan, and the 2011 Cyclone Yasi in Australia, to mention only the most globally visible of these events. Only a handful of these catastrophes brought about persistent global attention, and even fewer of them were recipients of the cultural heritage anxiety that is expected of heritage preservation organizations.

I saw each of these events as opportunities in which the place and role of heritage in postdisaster settings could be contextualized and better understood, where the aspirations put forward in a post-tsunami tourist leaflet could be substantiated to become a legitimate claim, taking advantage of each opportunity to address the permanence and impermanence of material culture and its preservation as situated constructions with specific roles in the future of each society. When the devastating Hurricane Katrina ravaged the south of the United States, soon after the Indian Ocean tsunami catastrophe, US/ICOMOS issued a written statement that expressed concern over the lack of establishment of a cohesive program of survey, damage recording, and assessment, and emergency stabilization for historic resources in the area. The organization urged the world's cultural and conservation community to assist in these efforts, especially those with experience in heritage recovery from catastrophic events.[18] However, more than five years later, when the earthquake devastated Haiti on January 12, 2010, international organizations appeared to be back at square one in their attempts at understanding and tackling the problem of patrimonial destruction in the region. It was apparent that no advances had been made toward identifying and addressing the challenges and potential outcomes of this particular context of destruction. What continued to rule in postdisaster contexts was the centrality of safeguarding at all costs and deploying the usual methodologies with a focus on already-identified heritage places. In this instance, once again, ICOMOS was flooded with offers of help from its members, reflecting the undeniable attraction that postdisaster rescue still entails. On January 14, 2010, the Blue Shield circulated a declaration that opened with its catchphrase: *culture is a basic need*, echoing once again the concept that "cultural heritage is a symbolic necessity . . . fundamental in rebuilding the identity, the dignity and the hope of the communities after a catastrophe."[19] In their January 2010 installment of this statement, the organization added that it placed the expertise and network of its member organizations at the disposal of their Haitian colleagues, calling on the international community, responsible authorities, and local population to fully support the efforts underway to preserve the rich and unique heritage of Haiti.

Without a field methodology proposed to accomplish this task or suggestions about the composition and training that a team of experts should represent, logistics for this type of aid deployment remained vague and their rhetoric superficial outside of the network of experts. Dozen of ICOMOS National Committees communicated their readiness to be deployed to the disaster area, while many of the twenty-eight scientific committees of ICOMOS mobilized their expertise to identify the most suitable candidates.[20] CyArk and the Kacyra Foundation in California offered their laser scanning equipment for documentation. A number of US universities put together groups of volunteers. A committee of experts was appointed to coordinate efforts, under the leadership of Dinu Bumbaru, likely the most active and expert voice in the ICOMOS leadership on matters of postconflict and postdisaster heritage preservation. I was part of the expert task force that was formed through this initiative. The main concern in the communications within the team was the development of an assessment methodology.[21] There were no existing methodologies in place, although this was not the first time that ICOMOS volunteered itself to be a provider of expertise in times of crisis. The few examples of rapid assessment of cultural heritage that we could have drawn inspiration from were, among others, the one-page short form used in post-Katrina New Orleans, and the fourteen-page long form that was proposed for postwar Iraq by the Getty Conservation Institute. In March, members of the steering committee began circulating a proposed methodology, a practical one-page assessment of condition and significance.[22] A second draft of the said methodology already extended from a one-page assessment to a six-part, eighteen-page document that delved further into the history of Haiti and the technical description of the earthquake before addressing the methodology itself.[23] This document reflected a tradition of preservation that places a high value on the resource as a historical asset, and echoes the knowledge-based model used in World Heritage List nomination processes. A fourth and final draft of the methodology, at twenty-seven pages, spent considerable effort presenting existing images, maps, and figures, including the location of Haiti on a map.[24]

Two conclusions could therefore be drawn from these institutional engagements related to natural disasters and heritage concerns over this brief period. From an institutional perspective, there remains a reluctance to commit resources to seriously improve preparation. But from the perspective of the debates proposed in the present volume, a broader and localized perspective is still lacking. Meanwhile, risk continues to be used as a rhetorical device for the safeguarding mission of heritage preservation but in unproductive ways that continue to marginalize methodologies. Instead of addressing cultural vulnerability in the present or acknowledging that what is at risk in postdisaster contexts is the cultural integrity of a people, this risk-centered perspective at play reiterates a message that is largely antiquated: heritage is about the past, not the present. In line with this bias, following Haiti, offers to apply new technologies for the documentation of cultural heritage places flooded communications within the expert task force. Without a clear output for these impressive imaging techniques, these efforts demonstrated how quickly preservation interests veer toward technocratic solutions to safeguard materiality, away from the social milieu in which postdisaster heritage can make an impact as a unique, temporally and spatially situated instrument of recovery.

As the tenth anniversary of the 2004 tsunami approached in late 2014, I was eager to document the place of heritage construction, now relatively far from the "aftermath" of destruction, from a retrospective position that acknowledged my temporal and spatial distance from the time of intensive fieldwork. The circulation of images in the Western media around that time stressed "before and after" landscapes, juxtaposing rubble and order, death and life.[25]

Consisting mostly of a commentary on resilience in the form of photo essays, the heritage "type" that was featured most prominently in this outlook was the Acehnese mosque and its heroic role in the disaster and its aftermath.[26] Another story that was circulated in the narrative of reordering was the reuniting of an Indonesian girl with her parents ten years later.[27] Disorder was also featured in the abandonment associated with the work of reconstruction that persisted since the early days of the recovery period, an abandonment that raised again the question of when recovery can be declared to be completed. On this note, a reporter for the UK newspaper the *Guardian* made the same contrasting observation that had been made in the years after the earliest reconstruction efforts, noting that "in the small seaside village of Lampu'uk—where the gigantic mosque was the only building that survived the tsunami—hundreds of houses donated by Turkish Red Cross are unoccupied."[28] The tsunami ruin persisted in many forms.

Ten years later, the spiritual interpretation of causality in this disaster continued to be associated with the difficulty of establishing an effective culture of prevention, arguing that the latter necessitates instead an embrace of a scientific rationale for the large number of casualties and damage already experienced.[29] Besides any suspected technological failings, what is being singled out in these critiques is often the cultural context in which these solutions are asked to take root. Surprisingly, a *National Geographic* article on this topic opened with an image of the tsunami ship PLDT Apung 1, but no commentary was made on the role that this type of heritage could have in supporting resilience and transmitting a lifesaving message that conforms and corresponds with a culture of disaster. The *Japan Times* described the variety of commemoration devices that can be found in Aceh, including the tsunami boats, featured as prominently as the lifesaving mosques.[30] In fact, ceremonies with religious overtones were taking place at the tsunami boats across Southeast Asia with the same formality that they had enjoyed in previous years, bringing light to the continuous relevance of these unlikely heritage monuments for the educational role that they can have, in tune with appropriate religious and scientific contexts.[31] In Aceh, photos of the ten-year anniversary displayed further efforts toward formalizing the Lampulo tsunami boat as a memorial, with improved signage and structures around it that distinguish it from an otherwise dislodged ruin.[32] Still unrecognized by heritage organizations that concern themselves with disaster preparedness, this construction of a heritage of disaster is ongoing and pivotal for meeting the objectives that heritage professionals align with—that is, protecting and enhancing the role of heritage so that, as it is aptly articulated in the quote earlier, "disastrous impacts can be minimized in the future and people can grow wiser."

Notes

1. With thanks to Byrn Williams and Barb Voss for this identification.
2. For example, paleolithic-style cultures: a Veddoid "flake culture" in central Sumatra and a more widespread Papua-Melanesoid "hand-axe culture" in the northeastern coast between Medan and Llokseumawe. According to Heine-Geldern's overview (1972), the latter is related to the Hoabinhian (c. 6000 BC) and early Bacsonien cultures of northeast Indochina, and cultures of Siam and the Malay Peninsula.
3. The reported figure of casualties associated with the tsunamis on December 26, 2004, and March 28, 2005, combined was a total of seven people.
4. A year after the tsunami, the human landscape of reconstruction was reported to involve 124 International NGOs plus 430 local NGOs who were identified as working in the reconstruction efforts, as well as dozens of donors, United Nations agencies, government agencies, and others.
5. "Aceh and Nias One Year after the Tsunami: A Joint Report of The BRR and International Partners," December 2005.
6. "Aceh and Nias One Year after the Tsunami: A Joint Report of The BRR and International Partners," December 2005, p. 4.
7. The director of the Aceh Museum, Dr. Nurdin, had recently stated that the content of the museum had not been discussed.
8. This project was supported by the Asia Research Institute of the National University of Singapore (ARI-NUS), supervised by Dr. Patrick Daly, who coordinated a twelve-member surveying team from AHCF, led by Yenny Rahmayati.
9. The BP3 reportedly spent funds in the post-tsunami period rehabilitating numerous heritage sites in and around the city. However, regional experts have commented that little if any was well spent, and most of these funds simply "disappeared." No attempt by the governmental institution responsible for this type of heritage work

was made to create a post-tsunami inventory of heritage monuments of any kind or involve the local NGOs and/or communities in Aceh.

10. Available at http://archive.cyark.org/hazard-map.
11. He also carried out the first systematic restoration project of the *candi* of Prambanan Temple. This project was continued under a new director and, with funds from the Netherlands, was completed in 1938. Restoration of ancient monuments from the Islamic period was also undertaken in projects such as the Gapura gravesite complex at Sendang Duwur between 1938 and 1940, the restoration of the Islamic gravesites in Arosbaya Bangkalan Madura, and the repair of Islamic monuments in Cirebon, such as the gravesite of Sunan Gunung Jati in 1938, the Sunyaragi Caves in 1939, and the Panjunan Mosque between 1939 and 1940. Colonial buildings began to be restored around this time, with the restoration of the city walls and old fort in Fort Rotterdam in Ujung Pandang between 1937 and 1938, and the sultan's palace at Yogyakarta in 1938. At the time of Indonesian independence, there were two offices for the antiquities institute, one in Jakarta and one in Prambanan.
12. The institute was divided into two directorates: the Directorate of Antiquities, which handled restorations, and the National Center for Archaeological Research.
13. However, during most of the time I researched Aceh (2006–2011), tourists were officially restricted from the region, as the embassies of Indonesia refused to issue entry visas to this area.
14. "Kebudayaan Aceh—Warisan Budaya Melayu Dunia," 2009 tourist pamphlet.
15. Image can be viewed here: http://rsd.gsfc.nasa.gov/rsd/images/Floyd.html.
16. The Maimun Saleh Airport is the current airport name and code for Sabang in Pulau Weh.
17. With special thanks to Dr. Nicholas Stanley-Price, who inadvertently drew my attention to the idea of "airmindedness" during the preparation of his volume *Imperial Outpost in the Gulf* (2012).
18. Message from US/ICOMOS executive director G. Araoz, dated September 23, 2005.
19. Haiti Blue Shield Statement, ICOMOS International Secretariat, email dated January 14, 2010, p. 1.
20. Email from ICOMOS, dated January 28, 2010.
21. Email from ICOMOS president Araoz to Haiti Steering Committee, dated January 12, 2010.
22. Email from Haiti Steering Committee, dated March 4, 2010.
23. Email from Haiti Steering Committee, dated March 8, 2010.
24. Email from Haiti Steering Committee, dated April 14, 2010.
25. For example, Reuters's slideshow "Tsunami Then and Now" (www.reuters.com/news/picture/tsunami-then-and-now?articleId=USRTR28AH2); the *New York Times*'s slideshow "Striking Before & After Shots Show Impact of Boxing Day Tsunami 10 Days Later" (www.nydailynews.com/news/boxing-day-tsunami-10-years-gallery-1.2051080); and *National Geographic*'s "Indonesia Tsunami Pictures: Banda Aceh, Then and Now" (http://news.nationalgeographic.com/news/2012/04/pictures/120412-indonesia-tsunami-earthquake-science-world/).

26. Photos of mosques are prominent in the aforementioned articles, but there were also dedicated notes to surviving mosques and the role they played in the aftermath of the disaster, in the *Guardian*'s "Banda Aceh: Where Community Spirit Has Gone but Peace Has Lasted" (www.theguardian.com/cities/2014/jan/27/banda-aceh-community-spirit-peace-indonesia-tsunami); *Euronews*'s "Indonesia Remembers Victims in Aceh, Area Worst Hit by 2004 Tsunami" (www.euronews.com/2014/12/26/indonesia-remembers-victims-in-aceh-area-worst-hit-by-2004-tsunami/); and the *Japan Times*'s "Remembering the 2004 Tsunami in Banda Aceh" (www.japantimes.co.jp/news/2014/12/21/asia-pacific/remembering-2004-tsunami-banda-aceh/#.VL4k-yuUd1A).
27. Featured in the *Strait Times*, "Indonesian Girl Swept Away by 2004 Tsunami Is Reunited with Parents" (www.straitstimes.com/news/asia/south-east-asia/story/indonesian-girl-swept-away-2004-tsunami-reunited-parents-20140808).
28. See the *Guardian*'s "Banda Aceh: Where Community Spirit Has Gone but Peace Has Lasted" (www.theguardian.com/cities/2014/jan/27/banda-aceh-community-spirit-peace-indonesia-tsunami).
29. For example, in *National Geographic*'s "Will Indonesia Be Ready for the Next Tsunami?" (http://news.nationalgeographic.com/news/2014/12/141226-tsunami-indonesia-catastrophe-banda-aceh-warning-science/); the *International Business Times*'s "2004 Indian Ocean Tsunami 10 Years Later: Warning System Installed After Disaster Has 'Critical Gaps'" (www.ibtimes.com/2004-indian-ocean-tsunami-10-years-later-warning-system-installed-after-disaster-has-1763662); and Reuters's "Asia Remembers Devastating 2004 Tsunami with Tears and Prayers" (www.reuters.com/article/2014/12/26/us-tsunami-anniversary-idUSKBN0K30PR20141226).
30. The *Japan Times* reported on December 21, 2014, that "One of the most visited sites is a long fishing boat that crashed on top of a house. A ramp leads to the roof, and you can also walk underneath where it is wedged between two dwellings. The boat provided a refuge for 56 survivors" (www.japantimes.co.jp/news/2014/12/21/asia-pacific/remembering-2004-tsunami-banda-aceh/#.VY7FUeuf_AO).
31. *BBC News Magazine*, December 24, 2014, "The Boat That Landed on a Roof and Saved 59 People" (www.bbc.com/news/magazine-30542317).
32. Australian Broadcasting Corporation, December 24, 2014, "Aceh—Fishing Boat on Roof of House" (www.abc.net.au/news/2014–12–24/aceh—fishing-boat-on-roof/5988106).

References

Abu El-Haj, Nadia. 2001. *Facts of the Ground: Archaeological Practice and Territorial Self-Fashioning in Israeli Society*. Chicago: Chicago University Press.

Adams, Cassandra. 1998. Japan's Ise Shrine and Its Thirteen-Hundred-Year-Old Reconstruction Tradition. *Journal of Architectural Education* 52 (1): 49–60.

Adams, Kathleen M. 2001. Danger-Zone Tourism: Prospects and Problems for Tourism in Tumultuous Times. In *Interconnected Worlds: Tourism in Southeast Asia*, edited by Peggy Teo, T. C. Chang, and K. C. Ho, 265–281. Amsterdam: Pergamon.

Affandy, Frances. 1999. Bandung Society for Heritage Conservation. In *Monuments and Sites Indonesia*, edited by Frances B. Affandy and Ahmad Rida Soemardi, 186–191. Bandung: PF Book, ICOMOS Indonesia and the Bandung Society for Heritage Conservation (Paguyuban Pelestarian Budaya Bandung).

Affandy, Frances B., and Ahmad Rida Soemardi, eds. 1999. *Monuments and Sites Indonesia*. Bandung: PF Book, ICOMOS Indonesia and the Bandung Society for Heritage Conservation (Paguyuban Pelestarian Budaya Bandung).

Alexander, Jeffery, Ron Eyerman, Bernard Giesen, Neil J. Smelser, and Piotr Sztompka, eds. 2004. *Cultural Trauma and Collective Identity.* Berkeley: University of California Press.

Ambary, Hasan Muarif. 1999. General Overview of Archeology in Indonesia. In *Monuments and Sites Indonesia*, edited by Frances B. Affandy and Ahmad Rida Soemardi, 9–16. Bandung: PF Book, ICOMOS Indonesia and the Bandung Society for Heritage Conservation (Paguyuban Pelestarian Budaya Bandung).

Andaya, Leonard Y. 1992. Interactions with the Outside World and Adaptation in Southeast Asian Society, 1500–1800. In *From Early Times to C. 1800*. Vol. 1 of *The Cambridge History of Southeast Asia*, edited by Nicolas Tarling, 345–401. Cambridge, UK: Cambridge University Press.

Anderson, Benedict. 1993. *Imagined Communities: Reflections on the Origin and Spread of Nationalism*. New York: Verson.

Anderson, Benedict. 1999. Indonesian Nationalism Today and in the Future. *Indonesia* 67: 1–12.

Appadurai, Arjun. 1996. *Modernity at Large: Cultural Dimensions of Globalization.* Minneapolis: University of Minnesota Press.

Appiah, Kwame Anthony. 2006. *Cosmopolitanism: Ethics in a World of Strangers*. New York: Norton.

Arif, Kamal. 2008. *Ragam Citra Kota: Banda Aceh: Interpretasi Sejarah, Memori Kolektif dan Arketipe Arsitekturnya*. Bandar Baru, Banda Aceh: Pustaka Bustanussalatin.

Aspinall, Edward. 2008. *Peace without Justice? The Helsinki Peace Process in Aceh*. Geneva: Center for Humanitarian Dialogue.

Australia ICOMOS. 1999. *The Burra Charter: The Australia ICOMOS Charter for the Conservation of Places of Cultural Significance*. Burwood, Australia: Deakin University.

Baehr, Peter. 2008. Colonialism, Slavery, and the Slave Trade: A Dutch Perspective. In *The Age of Apology: Facing Up to the Past*, edited by Mark Gibney, Rhoda E. Howard-Hassmann, Jean-Marc Coicaud, and Niklaus Steiner, 229–240. Philadelphia: University of Pennsylvania Press.

Baillie, Britt, and Christopher Chippindale. 2007. Tangible-Intangible Cultural Heritage: A Sustainable Dichotomy? The 7th Annual Cambridge Heritage Seminar. McDonald Institute for Archaeological Research, University of Cambridge. *Conservation and Management of Archaeological Sites* 8 (3): 174–176.

Ballard, Chris. 2000. The Fire Next Time: The Conversion of the Huli Apocalypse. *Ethnohistory* 47 (1): 205–225.

Baudrillard, Jean. 1994. *Simulacra and Simulation*. Ann Arbor: University of Michigan Press

Beck, Ulrich. 1992. *Risk Society: Towards a New Modernity*. London: SAGE.

Beck, Ulrich. 2002. The Terrorist Threat: World Risk Society Revisited. *Theory, Culture and Society* 19 (4): 39–55.

Benarie, Michel, and James R. Druzik. 1992. Catastrophes and Entropy of Cultural Properties. In *Material Issues in Art and Archaeology III*, edited by Pamela B. Vandiver and James R. Druzik, 17–28. Materials Research Society Symposium Proceedings, vol. 267. Pittsburgh, PA: Materials Research Society.

Bille, Mikkel, Frida Hastrup, and Tim F. Sørensen, eds. 2010. *An Anthropology of Absence: Materializations of Transcendence and Loss*. Berlin: Springer.

Blake, Janet. 2000. On Defining Cultural Heritage. *The International and Comparative Law Quarterly* 49 (1): 61–85.

Blong, Russell J. 1982. *The Time of Darkness: Local Legends and Volcanic Reality in Papua New Guinea*. Seattle: University of Washington Press.

Bortolotto, Chiara. 2007. From the "Monumental" to the "Living" Heritage: A Shift in Perspective. In *World Heritage: Global Challenges, Local Solutions: Proceedings of a Conference at Coalbrookdale, 4–7 May 2006, hosted by the Ironbridge Institute*, edited by Roger White and John Carman, 39–45. BAR International Series, vol. 1698. Oxford: Archaeopress.

Bouchard, Chad. 2007. Tsunami Museum Sparks Debate in Indonesia. *Voice of America: Jakarta*. August 27.

Bouwsema-Raap, Wilhelmina. 2009. *The Great Mosque of Banda Aceh: Its History, Architecture and Relationship to the Development of Islam in Sumatra*. Bangkok, Thailand: White Lotus Press.

Breglia, Lisa. 2006. *Monumental Ambivalence: The Politics of Heritage*. Austin: University of Texas Press.
Buchli, Victor. 2002. *The Material Culture Reader*. Oxford: Berg.
Buchli, Victor. 2015. *An Archaeology of the Immaterial*. New York: Routledge.
Buchli, Victor, and Gavin Lucas. 2001. Presencing Absence. In *Archaeologies of the Contemporary Past*, edited by Victor Buchli and Gavin Lucas, 171–174. London: Routledge.
Burke, Peter. 1969. *The Renaissance Sense of the Past*. New York: St. Martin's Press.
Byrne, Denis. 1991. Western Hegemony in Archaeological Heritage Management. *History and Anthropology* 5 (2): 269–276.
Byrne, Denis. 1995. Buddhist Stuppa and Thai Social Practice. *World Archaeology* 27 (2): 266–281.
Byrne, Denis. 1997. The Archaeology of Disaster. *Public History Review* 6: 17–29.
Byrne, Denis. 2004. Chartering Heritage in Asia's Postmodern World. *Conservation: The Getty Conservation Institute Newsletter* 19 (2): 16–19.
Byrne, Denis. 2007. *Surface Collection: Archaeological Travels in Southeast Asia*. Lanham, MD: Altamira Press.
Byrne, Denis. 2014. *Counterheritage: Critical Perspectives on Heritage Conservation in Asia*. London: Routledge.
Byrne, Denis, Helen Brayshaw, and Tracy Ireland. 2003. *Social Significance: A Discussion Paper*. Hurtsville, Australia: National Parks and Wildlife Service.
Byrne, Denis, and Maria Nugent. 2004. *Mapping Attachment: A Spatial Approach to Aboriginal Post-Contact Heritage*. Hurtsville, Australia: Department of Environment and Conservation.
Cameron, Fiona. 2008. The Politics of Heritage Authorship: The Case of Digital Heritage Collections. In *New Heritage: New Media and Cultural Heritage*, edited by Yehuda Kalay, Thomas Kvan, and Janice Affleck, 170–184. London: Routledge.
Carman, John, and Marie Louise Stig Sørensen. 2012. Heritage Studies: An Outline. In *Heritage Studies: Methods and Approaches*, edited by Marie Louise Stig Sørensen and John Carman, 11–28. London: Routledge.
Castañeda, Quetzil E., and Christopher N. Matthews. 2008. Introduction: Ethnography and the Social Construction of Archaeology. In *Ethnographic Archaeologies*, edited by Quetzil E. Castañeda and Christopher N. Matthews, 1–23. Lanham, MD: Altamira Press.
Chadha, Ashish. 2006. Ambivalent Heritage: Between Affect and Ideology in a Colonial Cemetery. *Journal of Material Culture* 11 (3): 339–363.
Chamberlin, Eric Russell. 1979. *Preserving the Past*. London: Dent.
Chong, Jinn Winn. 2012. Mine, Yours or Ours? The Indonesia-Malaysia Disputes over Shared Cultural Heritage. *Sojourn: Journal of Social Issues in Southeast Asia* 27 (1): 1–53.
Clark, Marshall, and Juliet Pietsch, eds. 2014. *Indonesia-Malaysia Relations: Cultural Heritage, Politics and Labour Migration*. New York: Routledge.
Clavir, Miriam Lisa. 2001. The Future of Ethnographic Conservation: A Canadian Perspective. In *The British Museum Occasional Paper*, vol. 145, edited by Andrew Oddy and Sandra Smith, 57–60. London: British Museum.

Clavir, Miriam Lisa. 2002. *Preserving What Is Valued: Museums, Conservation and First Nations*. Vancouver: UBC Press.

Cleere, Henry. 2000. The World Heritage Convention in the Third World. In *Cultural Resource Management in Contemporary Society: Perspectives on Managing and Presenting the Past*, edited by Alf Hatton and Francis P. MacManamon, 99–106. London: Routledge.

Cohen, Mitchell. 1992. Rooted Cosmopolitanism. *Dissent* Fall: 478–483.

Colwell-Chanthaphonh, Chip, and T. J. Ferguson. 2006. Rethinking Abandonment in Archaeological Contexts. *The SAA Archaeological Record* 6 (1): 37–41.

Connerton, Paul. 1989. *How Societies Remember*. Cambridge, UK: Cambridge University Press.

Cortesão, Armando, Pires Tome, and Francisco Rodrigues, eds. 1944. *The Suma Oriental of Tome Pires: An Account of the East, from the Red Sea to Japan, Written in Malacca and India in 1512–1515*. London: Hakluyt Society.

Coté, Joost. 2002. Searching for Semarang: Nation, Urban Memory, and Cultural Heritage. In *The Disappearing "Asian" City: Protecting Asia's Urban Heritage in a Globalizing World*, edited by William Stewart Logan, 124–141. New York: Oxford University Press.

Coté, Joost. 2009. Postcolonial Shame: Heritage and the Forgotten Pain of Civilian Women Internees in Java. In *Places of Pain and Shame: Dealing with "Difficult Heritage*," edited by William Stewart Logan and Keir Reeves, 128–143. London: Routledge.

Cowherd, Robert. 1999. Indonesia and the New Conservation. In *Monuments and Sites Indonesia*, edited by Frances B. Affandy and Ahmad Rida Soemardi, 1–8. Bandung: PF Book, ICOMOS Indonesia and the Bandung Society for Heritage Conservation (Paguyuban Pelestarian Budaya Bandung).

Cribb, Robert. 2002. Unresolved Problems in the Indonesian Killings of 1965–1966. *Asian Survey* 42 (4): 550–563.

Daly, Patrick, and Yenny Rahmayati. 2012. Cultural Heritage and Community Recovery in Post-Tsunami Aceh. In *From the Ground Up, Perspective on Post-Tsunami and Post-Conflict Aceh*, edited by Patrick Daly, Yenny Rahmayati, and R. Michael Feener, 57–78. Singapore: ISEAS.

Daly, Patrick, Yenny Rahmayati, and R. Michael Feener, eds. 2009. *From the Ground Up, Perspective on Post-Tsunami and Post-Conflict Aceh*. Singapore: ISEAS.

Damais, Soedarmadji. 1992. Pusaka in Times of Change. In *Pusaka: Art of Indonesia*, edited by Haryati Soebadio, 205–208. Singapore: Archipelago Press.

Damm, Charlotte. 2005. Archaeology, Ethno-history and Oral Traditions: Approaches to the Indigenous Past. *Norwegian Archaeological Review* 38 (2): 73–87.

Dann, Graham M. S., and A. V. Seaton, eds. 2002. *Slavery, Contested Heritage and Thanatourism*. London: Haworth Hospitality Press.

Darmaputera, Eka. 1988. *Pancasila and the Search for Identity and Modernity in Indonesian Society*. Leiden: Brill.

Davies, Hugh. 2002. Tsunamis and the Coastal Communities of Papua New Guinea. In *Natural Disasters and Cultural Change*, edited by R. J. G. Torrence, 28–42. London: Routledge.

DeSilvey, Caitlin, and Tim Edensor. 2013. Reckoning with Ruins. *Progress in Human Geography* 37 (4): 465–485.

de Vries, Gerrit, and Dorothee C. Segaar-Höweler. 2009. *Henri Maclaine Pont (1884–1971): Architect, Constructeur, Archeoloog.* Rotterdam: Stichting BONAS.

Dittmer, Lowell. 2002. The Legacy of Violence in Indonesia. *Asian Survey* 42 (4): 541–544.

Dominey-Howes, Dale, Phil Cummins, and David Burbidge. 2007. Historic Records of Teletsunami in the Indian Ocean and Insights from Numerical Modelling. *Natural Hazards* 42 (1): 1–17.

Douglas, Mary. 1966. *Purity and Danger: An Analysis of Concepts of Pollution and Taboo.* London: Routledge & Keagan Paul.

Douglas, Mary. 1992. *Risk and Blame: Essays in Cultural Theory.* London: Routledge.

Douglas, Mary, and Aaron Wildavsky. 1982. *Risk and Culture: An Essay on the Selection of Technical and Environmental Dangers.* Los Angeles: University of California Press.

Douglass, Ana, and Thomas Vogler. 2003. *Witness and Memory: The Discourse of Trauma.* New York: Routledge.

Drewes, Geradus Willebrordus Joannes, ed. 1979. *Hikajat Potjut Muhamat: An Acehnese Epic.* The Hague: Nijhoff.

Drexler, Elizabeth F. 2008. *Aceh, Indonesia: Securing the Insecure State (The Ethnography of Political Violence).* Philadelphia: University of Pennsylvania Press.

Druzik, James. 1992. Introduction to Part I: Cultural Heritage in Conflict. In *Material Issues in Art and Archaeology III*, edited by Pamela B. Vandiver and James R. Druzik, 1. Materials Research Society Symposium Proceedings, vol. 267. Pittsburgh, PA: Materials Research Society.

Dzikron, A. M. 2009. *Tragedi Tsunami di Aceh: Bencana Alam atau Rekayasa.* Solo: Muhammad Taufiq & Partners.

Ebron, Paulla. 2007. *Memory's Enactments: Texts, Monuments, Performance.* Stanford, CA: Stanford University.

Edmonds, Leigh. 1993. How Australians Were Made Airminded. *Continuum: The Australian Journal of Media & Culture* 7 (1): 183–206.

Engelhardt, Richard A., and Pamela R. Rogers. 2009. *Hoi An Protocols for Best Conservation Practice in Asia.* Bangkok: UNESCO Bangkok.

Engle Merry, Sally. 2006. Transnational Human Rights and Local Activism: Mapping the Middle. *American Anthropologist* 108 (1): 38–51.

Erni, Christian, ed. 2008. *The Concept of Indigenous Peoples in Asia: A Resource Book.* Copenhagen: International Work Group for Indigenous Affairs.

Ervin, Alexander W. 1997. Trying the Impossible: Relatively "Rapid" Methods in a City-Wide Needs Assessment. *Human Organization* 56 (4): 379–387.

Exell, Karen, and Trinidad Rico. 2014. *Cultural Heritage in the Arabian Peninsula: Debates, Discourse and Practices.* Surrey, UK: Ashgate.

Fealy, Greg. 2006. Islam in Southeast Asia. In *Southeast Asian History: Essential Readings*, edited by Damodar R. SarDesai, 305–323. Boulder, CO: Westview Press.

Feener, R. Michael. 2011. The Acehnese Past and Its Present State of Study. In *Mapping the Acehnese Past*, edited by Michael R. Feener, Patrick Daly, and Anthony Reid, 1–24. Leiden: KITLV.

Feener, R. Michael. 2014. *Shari'a and Social Engineering: The Implementation of Islamic Law in Contemporary Aceh, Indonesia*. Oxford: Oxford University Press.

Ferguson, Thomas John, and Roger Anyon. 2001. Hopi and Zuni Cultural Landscapes: Implications of History and Scale for Cultural Resources Management. In *Native Peoples of the Southwest: Negotiating Land, Water, and Ethnicities*, edited by Laura Lee Weinstein, 99–122. Westport: Bergin & Garvey.

Filippucci, Paola. 2009. Heritage and Methodology: A View from Social Anthropology. In *Heritage Studies: Methods and Approaches*, edited by Marie Louise Stig Sørensen and John Carman, 319–325. London: Routledge.

Forty, Adrian, and Susanne Kuchler, eds. 1999. *The Art of Forgetting*. Oxford: Berg.

Foucault, Michel. 1970. *The Order of Things*. London: Routledge.

Fowles, Severin, and Kaet Heupel. 2013. Absence. In *The Oxford Handbook of the Archaeology of the Contemporary World*, edited by Paul Graves-Brown, Rodney Harrison, and Angela Piccini, 178–191. Oxford: Oxford University Press.

Frederick, William H., and Robert L. Worden. 1993. *Indonesia: A Country Study*. Washington, DC: Federal Research Division, Library of Congress.

Fuller Collins, E. 2002. Indonesia: A Violent Culture? *Asian Survey* 42 (4): 582–602.

Geertz, Clifford. 1968. *Islam Observed: Religious Development in Morocco and Indonesia*. Chicago: University of Chicago Press.

Gibb, Hamilton Alexander Rosskeen, ed. 1929. *Ibn Battuta: Travels in Asia and Africa, 1325–1354*. London: Routledge.

Gin, Ooi Keat, ed. 2004. *Southeast Asia: A Historical Encyclopaedia, from Angkor Wat to East Timor*. Santa Barbara, CA: ABC Clio.

Glover, Ian C. 2003. National and Political Uses of Archaeology in South-East Asia. *Indonesia and the Malay World* 31 (89): 16–30.

Gordillo, Gastón. 2014. *Rubble: The Afterlife of Destruction*. Durham, NC: Duke University Press.

Guillot, Claude, and Ludvik Kalus. 2008. *Inskripsi Islam tertua di Indonesia*. Kepustakaan Populer Gramedia: Ecole Francaise d'Extreme-Orient.

Haber, Wolfgang. 1995. Concept, Origin, and Meaning of Landscape. In *Cultural Landscapes of Universal Value: Components of a Global Strategy*, edited by Bernd von Droste, Harald Plachter, and Mechtild Rossler, 38–42. New York: UNESCO.

Hadi, Amirul. 2004. *Islam and State in Sumatra: A Study of Seventeenth-Century Aceh*. Leiden: Brill.

Halbwachs, Maurice, and Lewis A. Coser. 1992. *On Collective Memory*. Chicago: University of Chicago Press.

Hancock, Mary E. 2008. *The Politics of Heritage from Madras to Chennai*. Bloomington: Indiana University Press.

Haraway, Donna J. 1991. Situated Knowledges. In *Simians, Cyborgs, and Women: The Reinvention of Nature*, edited by Donna J. Haraway, 183–200. New York: Routledge.

Hardjono, Joan M. 1971. *Indonesia, Land and People*. Djakarta: Gunung Agung.

Harrison, Rodney. 2004. *Shared Landscapes: Archaeologies of Attachment to the Pastoral Industry in New South Wales*. Sydney: UNSW Press.

Harrison, Rodney. 2011. Surface Assemblages: Towards an Archaeology in and of the Present. *Archaeological Dialogues* 18 (2): 141–161.

Harrison, Rodney. 2013. *Heritage: Critical Approaches*. London: Routledge.

Harrison, Rodney. 2016. World Heritage Listing and the Globalisation of the Endangerment Sensibility. In *Endangerment, Biodiversity and Culture*, edited by Nelia Dias and Fernando Vidal, 195–217. London: Routledge.

Harrison, Rodney, and John Schofield. 2010. *After Modernity: Archaeological Approaches to the Contemporary Past.* Oxford: Oxford University Press.

Hartono, Dibyo. 1999. Indonesian Architecture in the Colonial Period. In *Monuments and Sites Indonesia*, edited by Frances B. Affandy and Ahmad Rida Soemardi, 75–80. Bandung: PF Book, ICOMOS Indonesia and the Bandung Society for Heritage Conservation (Paguyuban Pelestarian Budaya Bandung).

Harvey, David C. 2001. Heritage Pasts and Heritage Presents: Temporality, Meaning and the Scope of Heritage Studies. *International Journal of Heritage Studies* 7 (4): 319–338.

Heine-Geldern, Robert. 1972. The Archaeology and Art of Sumatra. In *Sumatra: Its History and People*, edited by Edwin M. Loeb and Robert Heine-Geldern, 305–331. London: Oxford University Press.

Herscher, Andrew. 2010. *Violence Taking Place: The Architecture of the Kosovo Conflict*. Stanford, CA: Stanford University Press.

Hetherington, Kevin. 2004. Second-Handedness: Consumption, Disposal and Absent Presence. *Environment and Planning D: Society and Space* 22 (1): 157–173.

Hitchcock, Michael, and Victor T. King. 2003. Discourses with the Past: Tourism and Heritage in South-East Asia. *Indonesia and the Malay World* 31 (89): 3–15.

Ho, Engseng. 2004. Empire through Diasporic Eyes: A View from the Other Boat. *Society for Comparative Study of Society and History* 4: 210–246.

Ho, Engseng. 2006. *The Graves of Tarim: Genealogy and Mobility across the Indian Ocean.* Berkeley: University of California Press.

Hoffman, Susanna, and Anthony Oliver-Smith, eds. 2002. *Catastrophe and Culture: The Anthropology of Disaster.* Santa Fe, NM: School of American Research Advanced Seminar Series.

Hollowell, Julie, and George Nicholas. 2009. Using Ethnographic Methods to Articulate Community-Based Conceptions of Cultural Heritage Management. *Public Archaeology: Archaeological Ethnographies* 8 (2–3): 141–160.

Holtorf, Cornelius. 2001. Is the Past a Non-renewable Resource? In *Destruction and Conservation of Cultural Property*, edited by Peter T. Layton, Robert Stone, and Julian Thomas, 286–297. London: Routledge.

Holtorf, Cornelius. 2013. On Pastness: A Reconsideration of Materiality in Archaeological Object Authenticity. *Anthropological Quarterly* 86 (2): 427–443.

Holtorf, Cornelius. 2015. Averting Loss Aversion in Cultural Heritage. *International Journal of Heritage Studies (IJHS)* 21: 405–421.

Holtorf, Cornelius, and Tim Schadla-Hall. 1999. Age as Artefact: On Archaeological Authenticity. *European Journal of Archaeology* 2 (2): 229–247.

Howe, Renate, and William S. Logan. 2002. Conclusion: Protecting Asia's Urban Heritage: The Way Forward. In *The Disappearing "Asian" City: Protecting Asia's*

Urban Heritage in a Globalizing World, edited by William S. Logan, 245–256. New York: Oxford University Press.

Hurgronje, C. Snouck. 1906. *The Acehnese*. Translated by A. W. S. O'Sullivan. Leiden: Brill.

Huyssen, Andreas. 2003. *Present Pasts: Urban Palimsests and the Politics of Memory*. Stanford, CA: Stanford University Press.

ICOMOS. 1964. *The Venice Charter: International Charter for the Conservation and Restoration of Monuments and Sites Second International Congress of Architects and Technicians of Historic Buildings*. Retrieved from www.icomos.org/charters/venice_e.pdf.

ICOMOS. 2004. *The World Heritage List: Filling the Gaps—An Action Plan for the Future*. International Council for Monuments and Sites. Retrieved from whc.unesco.org/document/102409.

ICOMOS Symposia. 1994. *The Nara Document on Authenticity*. Retrieved from whc.unesco.org/document/9379.

Iemura, Hirokazu M., and Mulyo Harris Pradono. 2007. Tsunami Height Memorial Poles in Banda Aceh for Disaster Prevention and Education. In *Proceedings of the International Symposium on Engineering Lessons Learned from the 2011 Great East Japan Earthquake, March 1–4, 2012, Tokyo, Japan*, 1605–1616. Tokyo: Japan Association for Earthquake Engineers.

Jackson, Andrew. 2008. Local and Regional History as Heritage: The Heritage Process and Conceptualising the Purpose and Practice of Local Historian. *International Journal of Heritage Studies* 14 (4): 362–379.

Jauhola, Marjaana. 2013. *Post-Tsunami Reconstruction in Indonesia: Negotiating Normativity through Gender Mainstreaming Initiatives in Aceh*. London: Routledge.

Jigyasu, Rohit. 2014. Personal communication. December 4. Doha, Qatar.

Jones, Barclay G., ed. 1986. *Protecting Historic Architecture and Museum Collections from Natural Disasters*. Stoneham, MA: Butterworths.

Jones, Roy, and Brian J. Shaw. 2006. Palimpsests of Progress: Erasing the Past and Rewriting the Future in Developing Societies—Case Studies of Singapore and Jakarta. *International Journal of Heritage Studies* 12 (2): 122–138.

Joy, Charlotte. 2012. *The Politics of Heritage Management in Mali: From UNESCO to Djenne*. Walnut Creek, CA: Left Coast Press.

Karlström, Anna. 2005. Spiritual Materiality: Heritage Preservation in a Buddhist World? *Journal of Social Archaeology* 5 (3): 338–355.

Karlström, Anna. 2009. *Preserving Impermanence: The Creation of Heritage in Vientiane, Laos*. Uppsala: Institutionen för arkeologi och antik historia.

Kartiwa, Suwati. 1992. Pusaka and the Palaces of Java. In *Pusaka: Art of Indonesia*, edited by Haryati Soebadio, 159–164. Singapore: Archipelago Press.

Kartomi, Margaret J. 1981. Music and Dance in Aceh: A Preliminary Survey. *Indonesia Circle* 9 (24): 15–28

Khan, Sher Banu A. L. 2011. The Jewel Affair: The Sultana, Her Orang Kaya and the Dutch Foreign Envoys. In *Mapping the Acehnese Past*, edited by R. Michael Feener, Patrick Daly, and Anthony Reid, 141–162. Leiden: KITLV Press.

Kingsbury, Benedict. 1998. Indigenous Peoples in International Law: A Constructivist Approach to the Asian Controversy. *American Journal of International Law* 92: 414–457.

Kingsbury, Damien. 2006. *Peace in Aceh: A Personal Account of the Helsinki Peace Process*. Jakarta: Equinox.

Kreps, Christina F. 2003. *Liberating Culture: Cross-Cultural Perspectives on Museums, Curation, and Heritage Preservation*. London: Routledge.

Kreps, Christina F. 2005. Indigenous Curation as Intangible Cultural Heritage: Thoughts on the Relevance of the 2003 UNESCO Convention. *Theorizing Cultural Heritage* 1 (2): 3–8.

Kreps, Christina F. 2006. Non-Western Models of Museums and Curation in Cross-Cultural Perspective. In *A Companion to Museum Studies*, edited by Sharon Macdonald, 457–472. Oxford: Blackwell.

Küchler, Susanne. 1993. Landscape as Memory: The Mapping of Process and Its Representation in a Melanesian Society. In *Landscape Politics and Perspectives*, edited by Barbara Bender, 85–106. Oxford: Berg.

Kwon, Heonik. 2008. *Ghosts of War in Vietnam*. Studies in the Social and Cultural History of Modern Warfare. Cambridge, UK: Cambridge University Press.

Labadi, Sophia. 2005. A Review of the Global Strategy for a Balanced Representative and Credible World Heritage List 1994–2004. *Conservation and Management of Archaeological Sites* 7 (2): 89–102.

Lafrenz-Samuels, Kathryn, and Trinidad Rico, eds. 2015. *Heritage Keywords: Rhetoric and Redescription in Cultural Heritage*. Boulder: University Press of Colorado.

Lakoff, Andrew. 2007. Preparing for the Next Emergency. *Public Culture* 19 (2): 247–271.

Lakoff, George. 1990. *Women, Fire, and Dangerous Things*. Chicago: University of Chicago Press.

Larkham, Peter J. 1995. Heritage as Planned and Conserved. In *Heritage, Tourism and Society*, edited by David. T. Herbert, 85–116. London: Mansell.

Lay, Thorne, Hiroo Kanamori, Charles J. Ammon, Meredith Nettles, Steven N. Ward, Richard C. Aster, . . . and Stuart Sipkin. 2005. The Great Sumatra-Andaman Earthquake of 26 December 2004. *Science* 308 (5725): 1127–1133.

Layton, Robert, Peter G. Stone, and Julian Thomas, eds. 2001. *Destruction and Conservation of Cultural Property*. London: Routledge.

Layton, Robert, and Julian Thomas. 2001. Introduction: The Destruction and Conservation of Cultural Property. In *Destruction and Conservation of Cultural Property*, edited by Robert Layton, Peter G. Stone, and Julian Thomas, 1–21. London: Routledge.

Leach, Michael. 2009. Difficult Memories: The Independence Struggle as Cultural Heritage in East Timor. In *Places of Pain and Shame: Dealing with "Difficult Heritage,"* edited by William S. Logan and Keir Reeves, 144–161. London: Routledge.

Lennon, John, and Malcolm Foley. 2004. *Dark Tourism: The Attraction of Death and Disaster*. London: Continuum.

Leyden, John. 1821. *Malay Annals Translated from the Malay Language*. London: Longman, Hurst, Rees, Orme, and Brown.

Lipe, William D. 1984. Value and Meaning in Cultural Resources. In *Approaches to the Archaeological Heritage*, edited by Henry Cleere, 1–11. Cambridge, UK: Cambridge University Press.

Little, Barbara J., and Larry J. Zimmerman. 2010. In the Public Interest: Creating a More Activist, Civically-Engaged Archaeology. In *Voices in American Archaeology*, edited by Wendy Ashmore, Dorothy Lippert, and Barbara Mills, 131–159. Washington, DC: Society for American Archaeology Press.

Loeb, Edwin M. 1972. *Sumatra: Its History and People*. London: Oxford University Press.

Long, Colin, and Keir Reeves. 2009. "Dig a Hole and Bury the Past in It": Reconciliation and the Heritage of Genocide in Cambodia. In *Places of Pain and Shame: Dealing with "Difficult Heritage,"* edited by William Stewart Logan and Keir Reeves, 69–81. London: Routledge.

Low, Setha M. 1981. Social Science Methods in Landscape Architecture. *Landscape Planning* 8: 137–148.

Low, Setha M. 2002. Anthropological-Ethnographic Methods for the Assessment of Cultural Values in Heritage Conservation. In *Assessing the Values of Cultural Heritage*, edited by Marta de la Torre, 31–50. Los Angeles: The Getty Conservation Institute.

Lowenthal, David. 1985. *The Past is a Foreign Country*. Cambridge, UK: Cambridge University Press.

Lowenthal, David. 1998. *The Heritage Crusade and the Spoils of History*. Cambridge, UK: Cambridge University Press.

Lowenthal, David. 1999. Heritage Stewardship and the Amateur Tradition. *APT Bulletin* 30 (2/3): 7–9.

Macdonald, Sharon. 2009. Reassembling Nuremberg, Reassembling Heritage. *Journal of Cultural Economy* 2 (1–2): 117–134.

Marsden, William. 1783. *History of Sumatra: Containing an Account of the Government, Laws, Customs and Manners of the Native Inhabitants, with a Description of the Natural Productions, and a Relation to the Ancient Political State of the Island.* London: Payne.

Martinkus, John. 2004. *Indonesia's Secret War in Aceh*. Milsons Point: Random House Australia.

Mason, Randall, and Erica Avrami. 2002. Heritage Values and Challenges of Conservation Planning. In *Management Planning for Archaeological Sites: An International Workshop Organized by the Getty Conservation Institute*, edited by Jeanne Marie Teutonico and Gaetano Palumbo, 13–26. Los Angeles, CA: Getty Conservation Institute.

McAdoo, Brian G., Lori Dengler, Gegar Prasetya, and Vasily V. Titov. 2006. Smong: How an Oral History Saved Thousands on Indonesia's Simeulue Island during the December 2004 and March 2005 Tsunamis. *Earthquake Spectra* 22 (3): 661–669.

McKinnon, Edwards E. 1988. Beyond Serandib: A Note on Lambri at the Northern Tip of Aceh. *Indonesia* 46: 103–121.

Meskell, Lynn. 2002. Negative Heritage and Past Mastering in Archaeology. *Anthropological Quarterly* 75 (3): 557–574.

Meskell, Lynn. 2009. *Cosmopolitan Archaeologies*. Durham, NC: Duke University Press.

Meskell, Lynn. 2011. *The Nature of Heritage: The New South Africa*. Malden, MA: Wiley-Blackwell.

Meskell, Lynn. 2013. UNESCO's World Heritage Convention at 40 Challenging the Economic and Political Order of International Heritage Conservation. *Current Anthropology* 54 (4): 483–494.

Meskell, Lynn. 2014. States of Conservation: Protection, Politics, and Pacting within UNESCO's World Heritage Committee. *Anthropological Quarterly* 87 (1): 217–243.

Meyer, M. 2012. Placing and Tracing Absence: A Material Culture of the Immaterial. *Journal of Material Culture* 17 (1): 103–110.

Miller, Michelle Ann. 2009. *Rebellion and Reform in Indonesia: Jakarta's Security and Autonomy Policies in Aceh*. Oxon, UK: Routledge.

Miura, Keiko. 2005. Conservation of a "Living Heritage Site": A Contradiction in Terms? A Case Study of Angkor World Heritage Site. *Conservation and Management of Archaeological Sites* 7 (1): 3–18.

Miura, Keiko. 2007. A Note on the Current Impact of Tourism on Angkor and Its Environs. *Conservation and Management of Archaeological Sites* 8 (3): 132–135.

Monecke, Katrin, Willi Finger, David Klarer, Widjo Kongko, Brian G. McAdoo, Andrew L. Moore, and Sam U. Sudrajat. 2008. A 1000-Year Sediment Record of Tsunami Recurrence in Northern Sumatra. *Nature* 455: 1232–1234.

Morris, Eric E. 1985. Aceh: Social Revolution and the Islamic Vision. In *Regional Dynamics of the Indonesian Revolution: Unity from Diversity*, edited by Audrey R. Kahin, 83–110. Honolulu: University of Hawaii Press.

Morris-Suzuki, Tessa. 2005. *The Past within Us: Media, Memory, History*. New York: Verso.

Moses, John. 1992. Indigenous Perspectives on Conservation. *International Institute for Conservation, Canadian Group* 17 (3): 1–4.

Nalbant, Suleyman, Sandy Steacy, Kerry Sieh, Danny Natawidjaja, and John McCloskey. 2005. Seismology: Earthquake Risk on the Sunda Trench. *Nature* 435 (7043): 756–757.

Oliver-Smith, Anthony, and Susanna Hoffman, eds. 1999. *The Angry Earth: Disaster in Anthropological Perspective*. New York: Routledge.

Olsen, Bjørnar, and Þóra Pétursdóttir, eds. 2014. *Ruin Memories: Materialities, Aesthetics and the Archaeology of the Recent Past*. New York: Routledge.

Osborne, Hannah. 2014. 2004 Indian Ocean Earthquake and Tsunami: Facts about the Boxing Day Disaster. *International Business Times*, December 22. Retrieved from www.ibtimes.co.uk/2004-indian-ocean-earthquake-tsunami-facts-1480629.

Palmer, Scott W. 2006. *Dictatorship of the Air: Aviation Culture and the Fate of Modern Russia*. Cambridge Centennial of Flight. Cambridge, UK: Cambridge University Press.

Pannell, Sandra. 2007. Of Gods and Monsters: Indigenous Sea Cosmologies, Promiscuous Geographies and the Depths of Local Sovereignty. In *A World of Water: Rain, Rivers and Seas in Southeast Asian Histories*, edited by Peter Boomgaard, 71–102. Leiden: KITLV Press.

Peacock, James. 1973. *Indonesia: An Anthropological Perspective*. Pacific Palisades, UK: Goodyear.

Picard, Michel. 2003. The Touristification and Balinization in a Time of *Reformasi*. *Indonesia and the Malay World* 31 (89): 108–118.

Pombejra, Dhiravat Na. 2002. Towards a History of Seventeenth-Century Phuket. In *Recalling Local Pasts: Autonomous History in Southeast Asia*, edited by Sunait Chutintaranond and Chris Baker, 89–126. Chiang Mai: Silkworm Books

Purdey, Jemma. 2006. *Anti-Chinese Violence in Indonesia, 1996–1999*. Honolulu: University of Hawaii Press.

Pye, Elizabeth M. 2001. *Caring for the Past: Issues in Conservation for Archaeology and Museums*. London: James & James.

Qian, Fengqi. 2007. China's Burra Charter: The Formation and Implementation of the China Principles. *International Journal of Heritage Studies* 13 (3): 255–264.

Rachman, Erlita. 1999. Mass Media and the Protection of Monument and Sites. In *Monuments and Sites Indonesia*, edited by Frances B. Affandy and Ahmad Rida Soemardi, 192–196. Bandung: PF Book, ICOMOS Indonesia and the Bandung Society for Heritage Conservation (Paguyuban Pelestarian Budaya Bandung).

Reid, Anthony. 1979. *The Blood of the People: Revolution and the End of Traditional Rule in Northern Sumatra*. Kuala Lumpur, Malaysia: Oxford University Press.

Reid, Anthony. 1993. Islamization and Christianization in Southeast Asia: The Critical Phase, 1550–1650. In *Southeast Asia in the Early Modern Era: Trade, Power, and Belief*, edited by Anthony Reid, 151–179. Ithaca, NY: Cornell University Press

Reid, Anthony. 2005a. *An Indonesian Frontier: Acehnese and Other Histories of Sumatra*. Singapore: Singapore University Press.

Reid, Anthony. 2005b. Introduction: Sumatra as a Frontier. In *An Indonesian Frontier: Acehnese and Other Histories of Sumatra*, edited by Anthony Reid, 1–22. Singapore University Press, Singapore.

Reid, Anthony. 2005c. Remembering and Forgetting War and Revolution. In *Beginning to Remember: The Past in the Indonesian Present*, edited by Mary Sabina Zurbuchen, 168–232. Seattle: Singapore University Press in association with University of Washington Press.

Reid, Anthony. 2006. *The Verandah of Violence*. Singapore: Singapore University Press.

Reid, Anthony, and David Marr. 1979. *Perceptions of the Past in Southeast Asia*. Singapore: Heinemann Educational Books.

Rico, Trinidad. 2006. Conference Report: "Archaeology in Conflict: Cultural Heritage, Site Management and Sustainable Development in Conflict and Post-Conflict States in the Middle East," London, November 10–12. *Conservation and Management of Archaeological Sites* 8 (2): 105–107.

Rico, Trinidad. 2008. "Negative Heritage": The Place of Conflict in World Heritage. *Conservation and Management of Archaeological Sites* 10 (4): 344–352.

Rico, Trinidad. 2014. The Limits of a Heritage at Risk Framework: The Construction of Post-Disaster Cultural Heritage in Banda Aceh, Indonesia. *Journal of Social Archaeology* 14 (2): 157–176.

Rico, Trinidad. 2015. Heritage at Risk: The Authority and Autonomy of a Dominant Preservation Framework. In *Heritage Keywords: Rhetoric and Redescription in Cultural Heritage*, edited by Kathryn Lafrenz-Samuels and Trinidad Rico, 147–162. Boulder: University Press of Colorado.

Rico, Trinidad. Forthcoming. Technologies, Technocracy and the Promise of "Alternative" Heritage Values. In *Heritage in Action*, edited by Helaine Silverman, Emma Waterton, and Steve Watson. New York: Springer.

Ricoeur, Paul. 1990. *Time and Narrative*. Vol. 1. Chicago: University of Chicago Press.

Riddell, Peter G. 2006. Aceh in the Sixteenth and Seventeenth Centuries: "Serambi Mekkah" and Identity. In *The Verandah of Violence*, edited by Anthony Reid, 38–51. Singapore: Singapore University Press.

Riegl, Alois. 1996. The Modern Cult of Monuments: Its Essence and Its Development. In *Historical and Philosophical Issues in the Conservation of Cultural Heritage*, edited by Nicholas Stanley Price, M. Kirby Talley, Jr., and Alessandra Melucco Vaccaro, 69–83. Los Angeles, CA: Getty Conservation Institute.

Rojek, Chris. 1993. *Ways of Escape*. Basingstoke: Macmillan.

Rosaldo, Renato. 1980. *Ilongot Headhunting*. Stanford, CA: Stanford University Press.

Rossabi, Amy. 1999. Legal Policies surrounding the Degradation of Indonesia's Forests. *Vermont Journal of Environmental Law* 1: 29–49.

Ruskin, John. 1849. *The Seven Lamps of Architecture*. 4th ed. London: Smith, Elder.

Salazar, Noel B. 2014. Indonesia's World Heritage. In *Encyclopedia of Global Archaeology*, edited by Claire Smith, 3841–3846. New York: Springer.

Samuel, Raphael. 1994. *Theatres of Memory*. London: Verso.

SarDesai, Damodar R. 1989. *Southeast Asia: Past and Present*. San Francisco: Westview Press.

Seaton, A. V. 1996. Guided by the Dark: From Thanatopsis to Thanatourism. *International Journal of Heritage Studies* 2 (4): 234–244.

Severi, Carlo. 2001. Cosmology, Crisis, and Paradox: On the White Spirit in the Kuna Shamanistic Tradition. In *Disturbing Remains: Memory, History and Crisis in the Twentieth Century*, edited by Michael S. Roth and Charles G. Salas, 178–206. Los Angeles, CA: Getty Research Institute.

Shackel, Paul A. 2009. Foreword. In *Ethnographies and Archaeologies: Iterations of the Past*, edited by Lena Mortensen and Julie Hollowell, ix–x. Tampa: University Press of Florida.

Shanks, Michael. 1998. The Life of an Artifact in an Interpretive Archaeology. *Fennos Candia Archaeologica* 15: 15–20.

Shaw, Julia. 2000. Ayodhya's Sacred Landscape: Ritual Memory, Politics and Archaeological "Fact." *Antiquity* 74: 693–700.

Sherlock, Stephen. 2005. The Tyranny of Invented Traditions: Aceh. In *Violence in Between: Conflict and Security in Archipelagic Southeast Asia*, edited by Damien Kingsbury, 175–201. Clayton, Vic: Monash Asia Institute.

Siegel, James T. 1969. *The Rope of God*. Berkeley: University of California Press.

Siegel, James. 1979. *Shadow and Sound*. Chicago: Chicago University Press.

Siegel, James T., Joshua Barker, and Arief Djati. 2008. Notes of a Trip through Aceh, December 2007. *Indonesia* 86 (October): 1–54.

Sieh, Kerry, Patrick Daly, E. Edwards McKinnon, Jessica E. Pilarczyk, Hong-Wei Chiang, Benjamin Horton, Charles M. Rubin, Chuan-Chou Shen, Nazli Ismail, Christopher H. Vane, and R. Michael Feener. 2015. Penultimate Predecessors of the 2004 Indian Ocean Tsunami in Aceh, Sumatra: Stratigraphic, Archeological, and Historical Evidence. *Journal of Geophysical Research: Solid Earth* 120: 308–325.

Silverman, Helaine, and D. Fairchild Ruggles, eds. 2007. *Cultural Heritage and Human Rights*. New York: Springer.

Simamora, Adianto P. 2009. Govt to Seek Clarification on Pendet Dance with Malaysia. *The Jakarta Post,* August 23.

Sitwell, Christine, and Sarah Staniforth, eds. 1998. *Studies in the History of Painting Restoration*. London: Archetype and the National Trust.

Sixth International Congress of Architects. 1904. Recommendations of the Madrid Conference Sixth International Congress of Architects. Retrieved from www.getty.edu/conservation/publications_resources/research_resources/charters/charter01.html.

Smith, Holly S. 1998. *Aceh: Art and Culture.* Oxford: Oxford University Press.

Smith, Laurajane. 2006. *Uses of Heritage*. London: Routledge.

Society for the Protection of Ancient Buildings. 1877. The Manifesto of the Society for the Protection of Ancient Buildings. Society for the Protection of Ancient Buildings. Retrieved from www.spab.org.uk/what-is-spab/the-manifesto/.

Soebadio, Haryati. 1992. Introduction. In *Pusaka: Art of Indonesia*, edited by Haryati Soebadio, 15–16. Singapore: Archipelago Press.

Soekmono, R. 1965. Archaeology and Indonesian History. In *An Introduction to Indonesian Historiography*, edited by Soedjatmoko, 36–46. New York: Cornell University Press.

Soekmono, R. 1992. The Candi as Cultural Pusaka. In *Pusaka: Art of Indonesia,* edited by H. Soebadio, 55–66. Singapore: Archipelago Press.

Soemardi, Ahmad Rida. 1999. Heritage Conservation in Indonesia: Reflections and Future Issues. In *Monuments and Sites Indonesia*, edited by Frances B. Affandy and Ahmad Rida Soemardi, 197. Bandung: PF Book, ICOMOS Indonesia and the Bandung Society for Heritage Conservation (Paguyuban Pelestarian Budaya Bandung),

Sørensen, Marie L., and John Carman, eds. 2012. *Heritage Studies: Methods and Approaches*. London: Routledge.

Stanley-Price, Nicholas. (2012). *Imperial Outpost in the Gulf: The Airfield at Sharjah (UAE) 1932-52*. Philadelphia: Trans-Atlantic Publications.

Stoler, Ann Laura. 2002. On the Uses and Abuses of the Past in Indonesia: Beyond the Mass Killings of 1965. *Asian Survey* 42 (4): 650–642.

Stoler, Ann Laura. 2008. Imperial Debris: Reflections on Ruins and Ruination. *Cultural Anthropology* 23 (2): 191–219.

Stoler, Ann Laura. 2010. *Carnal Knowledge and Imperial Power: Race and the Intimate in Colonial Rule*. Berkeley: University of California Press.

Stoler, Ann, ed. 2013. *Imperial Debris: On Ruins and Ruination*. Durham, NC: Duke University Press.

Stovel, Herb. 1995. Monitoring World Cultural Heritage Sites. *ICOMOS Canada Bulletin* 4 (3): 15–20.

Stovel, Herb. 1998. *Risk Preparedness: A Management Manual for World Cultural Heritage*. Rome: ICCROM.
Stovel, Herb. 2007. Effective Use of Authenticity and Integrity as World Heritage Qualifying Conditions. *City & Time* 2 (3): 3.
Sturken, Marita. 2007. *Tourists of History: Memory, Kitsch, and Consumerism from Oklahoma City to Ground Zero.* Durham, NC: Duke University Press.
Sully, Dean, ed. 2008. *Decolonizing Conservation: Caring for Maori Meeting Houses outside New Zealand.* Walnut Creek, CA: Left Coast Press.
Tarlow, Peter E. 2005. Dark Tourism: The Appealing "Dark" Side of Tourism and More. In *Niche Tourism—Contemporary Issues, Trends and Cases*, edited by Marina Novelli, 47–58. Oxford: Elsevier.
Taylor, Jean Gelman. 2003. *Indonesia: Peoples and Histories*. New Haven, CT: Yale University Press.
Torrence, Robin, and John Grattan. 2002a. The Archaeology of Disasters: Past and Future Trends. In *Natural Disasters and Cultural Change*, Robin Torrence and John Grattan, 1–18. New York: Routledge.
Torrence, Robin, and John Grattan, eds. 2002b. *Natural Disasters and Cultural Change*. New York: Routledge.
Tsing, Anna L. 2005. *Friction: An Ethnography of Global Connection.* Princeton, NJ: Princeton University Press.
Turnbridge, J. E., and Gregory J. Ashworth. 1996. *Dissonant Heritage: The Management of the Past as a Resource in Conflict*. Chichester, UK: Wiley.
Turtinen, Jan. 2000. Globalising Heritage—On UNESCO and the Transnational Construction of a World Heritage. *SCORE Rapportserie* 2000: 12.
Vidal, Fernando, and Nélia Dias. 2016. Introduction: The Endangerment Sensibility. In *Endangerment, Biodiversity and Culture*, edited by Fernando Vidal and Nélia Dias, 1–38. Abingdon: Routledge.
UNESCO. 1954. Convention for the Protection of Cultural Property in the Event of Armed Conflict. Retrieved from http://portal.unesco.org/en/ev.php-URL_ID=13637&URL_DO=DO_TOPIC&URL_SECTION=201.html.
UNESCO. 1956. Recommendation on International Principles Applicable to Archaeological Excavation. Retrieved from http://portal.unesco.org/en/ev.php-URL_ID=13062&URL_DO=DO_TOPIC&URL_SECTION=201.html.
UNESCO. 1960. Campagne internationale pour la sauvegarde des monuments de la Nubie/International campaign to save the monuments of Nubia. *Museum* 13 (2): 65–66.
UNESCO. 1962. *Recommendation Concerning the Safeguarding of the Beauty and Character of Landscapes and Sites*. Retrieved from http://unesdoc.unesco.org/images/0016/001604/160477eb.pdf.
UNESCO. 1964. *Recommendation on the Means of Prohibiting and Preventing the Illicit Import, Export and Transfer of Ownership of Cultural Property*. Retrieved from http://unesdoc.unesco.org/images/0011/001145/114581f.pdf#page=149.
UNESCO. 1968. Recommendation Concerning the Preservation of Cultural Property Endangered by Public or Private Works UNESCO. Retrieved from http://

portal.unesco.org/en/ev.php-URL_ID=13085&URL_DO=DO_TOPIC&URL_SECTION=201.html.
UNESCO. 1972a. Convention Concerning the Protection of the World Cultural and Natural Heritage. Retrieved from http://whc.unesco.org/en/conventiontext/.
UNESCO. 1972b. *Operational Guidelines for the Implementation of the World Heritage Convention.* WHC. 12/01 Revised July 2012. Retrieved from http://whc.unesco.org/archive/opguide12-en.pdf
UNESCO. 2001. Proclamation of Masterpieces of the Oral and Intangible Heritage of Humanity. Retrieved from www.unesco.org/culture/ich/?pg=00103.
UNESCO. 2003. Convention for the Safeguarding of Intangible Cultural Heritage. Retrieved from www.unesco.org/culture/ich/en/convention.
UNESCO. 2007. World Heritage Center: Global Strategy. Retrieved from http://whc.unesco.org/en/globalstrategy/.
Valderrama, Fernando. 1995. *A History of UNESCO.* Paris: UNESCO.
Van Leur, Jacob Cornelius. 1955. *Indonesian Trade and Society.* The Hague: W. Van Hoeve.
Van Voorden, Frits. 1999. Foreword. In *Monuments and Sites Indonesia*, edited by Frances B. Affandy and Ahmad Rida Soemardi. Bandung: PF Book, ICOMOS Indonesia and the Bandung Society for Heritage Conservation (Paguyuban Pelestarian Budaya Bandung).
Viollet-le-Duc, Emile E. 1858–68. *Dictionnaire raisonné de l'architecture française du XIe au XVIe siècle*. Vol. 8. Paris: Bance.
Wall, Geoffrey, and Heather Black. 2004. Global Heritage and Local Problems: Some Examples from Indonesia. *Current Issues in Tourism* 7: 436–439.
Ward, Philip. 1975. *Indonesian Traditional Poetry.* Cambridge, UK: Oleander Press.
Warrack, Simon. 2011. Learning from Local Leaders: Working Together toward the Conservation of Living Heritage at Angkor Wat, Cambodia. *Change over Time* 1 (1): 34–51.
Wei, Chen, and Andreas Aass. 1989. Heritage Conservation: East and West. *ICOMOS Information* 3: 3–8.
West, Patrick. 2004. *Conspicuous Compassion: Why Sometimes it Really is Cruel to be Kind.* London: Civitas.
Wharton, Glenn. 2005. Indigenous Claims and Heritage Conservation: An Opportunity for Critical Dialogue. *Public Archaeology* 4: 199–204.
Whitehead, Anne. 2007. Introduction (Trauma). In *Theories of Memory: A Reader*, edited by Michael Rossington and Anne Whitehead, 186–191. Baltimore, MD: Johns Hopkins University Press.
Winter, Tim. 2007. *Post-Conflict Heritage, Postcolonial Tourism.* Oxon, UK: Routledge.
Winter, Tim, and Patrick Daly, eds. 2012. *The Routledge Handbook of Heritage in Asia.* London: Routledge.
Yamin. 1962. *Tatanegara Majapahit Sapta Parwa.* Jakarta: Prapantja.
Young, James E. 2003. Between History and Memory: The Voice of the Eyewitness. In *Witness and Memory: The Discourse of Trauma*, edited by Ana Douglass and Thomas Vogler, 275–283. New York: Routledge.

Yuill, Stephanie Marie. 2003. Dark Tourism: Understanding Visitor Motivation at Sites of Death and Disaster. Master's thesis, Texas A&M University.

Zilberg, Jonathan. 2008. The Acehnese Museum Conflict: Does Aceh Really Need a US$7.5 Million Tsunami Museum. Komunitas Tikar Pandan. Retrieved from www.tikarpandan.org/2009/07/the-acehnese-museum-conflict-does-aceh.html.

Zilberg, Jonathan. 2009. Memorials, State Domination and Inclusion versus Exclusion: The Case of the Tsunami Museum in Banda Aceh. *The International Journal of the Inclusive Museum* 2 (2): 99–110.

Zimmerman, Larry J. 2007. Plains Indians and Resistance to "Public" Heritage Commemoration of Their Pasts. In *Cultural Heritage and Human Rights*, edited by Helaine Silverman and D. Fairchild Ruggles, 144–158. New York: Springer.

Zinoman, Peter, and Nancy Lee Peluso. 2002. Rethinking Aspects of Political Violence in Twentieth-Century Indonesia and East Timor. *Asian Survey* 42 (4): 545–549.

Zurbuchen, Mary S. 2002. History, Memory, and the "1965 Incident" in Indonesia. *Asian Survey* 42 (4): 564–581.

Zurbuchen, Mary S. 2005. *Beginning to Remember: The Past in the Indonesian Present*. Seattle: University of Washington Press.

Index

For Product Safety Concerns and Information please contact our EU representative GPSR@taylorandfrancis.com Taylor & Francis Verlag GmbH, Kaufingerstraße 24, 80331 München, Germany